There Are Places in the World Where Rules Are Less Important Than Kindness

Carlo Rovelli is a theoretical physicist who has made significant contributions to the physics of space and time. He has worked in Italy and the US, and is currently directing the quantum gravity research group of the Centre de physique théorique in Marseille, France. His books *Seven Brief Lessons on Physics, Reality Is Not What It Seems,* and *The Order of Time* are international bestsellers which have been translated into forty-one languages.

There Are Places in the World Where Rules Are Less Important Than Kindness

CARLO ROVELLI

Translated by
Erica Segre and Simon Carnell

ALLEN LANE
an imprint of
PENGUIN BOOKS

ALLEN LANE

UK | USA | Canada | Ireland | Australia
India | New Zealand | South Africa

Allen Lane is part of the Penguin Random House group of companies
whose addresses can be found at global.penguinrandomhouse.com

First published by RCS MediaGroup S.p.A., Milan, 2018
This translation first published 2020
001

Copyright © Carlo Rovelli, 2018
Translation copyright © Erica Segre and Simon Carnell, 2020

The moral right of the author and of the translators has been asserted

Set in 12.5/15pt Garamond MT Std
Typeset by Jouve (UK), Milton Keynes
Printed and bound in Great Britain by Clays Ltd, Elcograf S.p.A.

A CIP catalogue record for this book is available from the British Library

ISBN: 978–0–241–45468–8

Contents

Preface

An article in a newspaper has something in common with a Japanese kōan or a European sonnet: limited in size and form, it can transmit little more than one piece of information, a single argument, one reflection, a single emotion. And yet it can speak about everything and anything.

The pieces collected here, which were published in various newspapers over the last decade, speak of poets, scientists and philosophers who have influenced me in some way, of travels, of my generation, of atheism, of black holes, telescopes, psychedelic experience, intellectual surprises . . . and much else. They are like brief diary entries recording the intellectual adventures of a physicist who is interested in many things and who is searching for new ideas – for a wide but coherent perspective.

The title has been borrowed from a phrase used in one of the articles: a phrase that perhaps conveys something of the spirit shared by these articles. Then again, perhaps it just reveals the spirit of the kind of world that I would like to live in . . .

Marseille, 2020

Aristotle the Scientist

Do objects of different weight fall at the same speed? At school we are told that, by letting balls drop from the Tower of Pisa, Galileo Galilei had demonstrated that the correct answer is yes. For the preceding two millennia, on the other hand, everyone had been blinded to the fact by the dogma of Aristotle, according to which the heavier the object, the faster it falls. Curiously, according to this story, it seems never to have occurred to anyone to test whether this was actually true or not before Francis Bacon and his contemporaries began observing nature and freed themselves from the straitjacket of Aristotelian dogmatism . . . It's a good story, but there's a problem with it. Try dropping a glass marble and a paper cup from a balcony. Contrary to what this beautiful story says, it is not at all true that they hit the ground at the same time: the heavier marble falls much faster, just as Aristotle says.

No doubt at this point someone will object that this happens because of air, the medium through which the things fall. True, but Aristotle did not write that things would fall at different speeds if we took out all the air. He wrote that things fall at different speeds in our world, where there is air. He was not wrong. He observed nature attentively. Better than generations of teachers and students who are prone to take things on trust, without testing them for themselves.

Aristotle's physics has had a lot of bad press. It has come to be thought of as built upon a priori assumptions, disengaged from observation, patently wrong-headed. This is substantially unjust. Aristotle's physics remained a reference point for Mediterranean civilization for so long not because it was dogmatic, but because it actually works. It provides a good description of reality, and a conceptual framework so effective that no one was able to better it for two thousand years. The essence of the theory is the idea that, in the absence of other influences, every object moves towards its 'natural place': lower down for earth, a little higher for water, higher again for air, and higher still for fire; the speed of 'natural movement' increases with weight and decreases according to the density of the medium in which the object is immersed. It's a simple, comprehensive theory that provides an elegant account of a great variety of phenomena – why smoke rises, for instance, and why a piece of wood drops down in air but floats upwards in water. As a theory it is obviously not perfect, but then we should remember that nothing in modern science is perfect either.

The bad reputation that has become attached to Aristotle's physics is partly the fault of Galileo, who in his writings launches a scathing all-out attack upon Aristotelian theory, portraying its adherents as fools. He did so for rhetorical reasons. But the bad reputation of Aristotle's physics is also due to the silly gulf that has opened up between scientific culture and humanist-philosophical discourse. Those who study Aristotle generally know little about physics, and those who are engaged in physics have little interest in Aristotle. The scientific brilliance of books by Aristotle such as his *On the Heavens* and *Physics* – the work from which the very discipline derives its name – is all too readily overlooked.

There is also another, more significant factor that explains our blindness to his scientific brilliance: the idea that it is impossible to compare the thought produced by cultural universes so distant from each other as those of Aristotle and of modern physics are, and that therefore we should not even try. Many historians today express horror at the idea of seeing Aristotelian physics as an approximation of Newtonian physics. In order to understand the original Aristotle, they argue, we must study him in the light of his context, and not through the conceptual frameworks of subsequent centuries. This may be true if we want to improve our understanding of Aristotle, but if we are interested in understanding today's knowledge, how it emerged from the past, it is precisely the relations between distant worlds that counts.

Philosophers and historians of science such as Karl Popper and Thomas Kuhn, who have had a strong influence on contemporary thought, have emphasized the importance of points of rupture in the course of the development of knowledge. Examples of such 'scientific revolutions', where an old theory is abandoned, include the move from Aristotle to Newton, and from Newton to Einstein. According to Kuhn, in the course of such passages a radical restructuring of thought takes place, to such a degree that the preceding ideas become irrelevant, incomprehensible even. They are 'incommensurable' with the subsequent theory, according to Kuhn. Popper and Kuhn deserve credit for having focused on this evolutive aspect of science and the importance of breaks, but their influence has also led to an absurd devaluation of the cumulative aspects of knowledge. Worse still is the failure to recognize the logical and historical relations between theories prior to and after every significant step forward. Newton's physics is perfectly recognizable as an approximation of Einstein's general relativity; Aristotle's theory is perfectly recognizable as an approximation contained within the theory of Newton.

This is not all, for within Newton's theory it is possible to recognize features of Aristotelian physics. For instance, the great idea of distinguishing the 'natural' motion of a body from that which has been 'forced' remains intact in Newtonian physics, as it does later in Einstein's theory. What changes is the role of gravity: it is the cause of forced motion in Newton (where natural motion is uniformly rectilinear), while it is an aspect of natural motion in Aristotle as well as, curiously, in Einstein (where natural motion, termed 'geodesic', returns to being that of an object in free fall, as in Aristotle). Scientists do not advance either as a result of mere accumulation of knowledge, or by means of absolute revolutions in which everything is thrown out and we begin again from zero. They advance instead, as in a wonderful analogy first made by Otto Neurath and frequently cited by Quine, 'like sailors who must rebuild their ship on the open sea, never able to start afresh from the bottom. Where a beam is taken away a new one must at once be put there, and for this the rest of the ship is used as support. In this way the ship can be shaped entirely anew, but only by gradual reconstruction.' In the great ship of modern physics we can still recognize its ancient structures – such as the distinction between natural and forced motion – as first laid out in the old ship of Aristotelian thought.

Let's go back to bodies falling through air or water and see what actually happens. The fall is neither at a constant speed and dependent on weight, as Aristotle maintained, nor at constant acceleration and independent of weight, as Galileo argued (not even if we ignore friction!). When an object falls, it goes through an initial stage during which it accelerates, then stabilizes at a constant speed which is greater for heavier bodies. This second stage is well described by Aristotle. The first stage, on the other hand, is usually very brief, difficult to observe, and as a result of this had escaped his notice. The existence of this initial stage had already been noted in antiquity: in the third century BC, for example, Strato of Lampsacus observed that a falling stream of water breaks into drops, indicating that the drops accelerate on falling, just like a line of traffic that breaks up as the vehicles accelerate.

In order to study this initial phase, which is difficult to observe because everything happens so quickly, Galileo devises a brilliant stratagem. Instead of observing falling bodies, he looks at balls rolling down a slight incline. His intuition, difficult to justify at the time but well founded, is that the 'rolling fall' of the balls reproduces that of bodies falling freely. In this way Galileo manages to record that at the beginning of the fall it is acceleration that remains constant, not speed. Galileo succeeded in uncovering the detail almost imperceptible to our senses where Aristotle's physics fails. It is like the observation used by Einstein at the beginning of the twentieth century in order to go beyond Newton: the movement of the planet Mercury, looked at closely, does not follow exactly the orbits calculated by Newton. In both cases, the devil is in the detail.

Einstein does to Newton what Galileo and Newton did to Aristotle: he shows that, for all its effectiveness, his version of physics is good only as a first approximation. Today we know that even Einstein's physics is not perfect: it fails when quantum physics enters into the equation. Einstein's physics needs to be improved upon as well. We are still not sure how.

Galileo did not build his new physics by rebelling against a dogma, or by forgetting Aristotle. On the contrary, having learned deeply from him, he worked out how to modify aspects of the Aristotelian conceptual cathedral: between himself and Aristotle there is not incommensurability but dialogue. I believe this is also the case at the borders between different cultures, individuals and peoples. It is not true, as today we love to repeat, that different cultural worlds are mutually impermeable and untranslatable. The opposite is true: the borders between theories, disciplines, eras, cultures, peoples and individuals are remarkably porous, and our knowledge is fed by the exchanges across this highly permeable spectrum. Our knowledge is the result of a continuous development of this dense web of exchanges. What interests us most is precisely this exchange: to compare, to exchange ideas, to learn and to build from difference. To mix, not to keep things separate.

There's quite some distance between Athens in the fourth century BC and seventeenth-century Florence. But there is no radical rupture, and no misunderstanding. It is because Galileo knows how to enter into dialogue with Aristotle, and to penetrate into the heart of his physics, that he finds the narrow opening through which it can be corrected and improved. He puts this beautifully himself, in a letter written in later life: 'I am certain that if Aristotle were to return to Earth he would receive me amongst his followers, in virtue of my very few contradictions of his doctrine.'

Lolita and the Blue Icarus

Passing through the Museum of Natural Sciences in Milan recently, I came across an old cabinet containing a collection of blue butterflies, together with what for me was an unexpected name in a context such as this: Vladimir Nabokov.

The same Nabokov, that is, who was the author of such dazzlingly written novels as *Lolita*:

> Lo-Lee-Ta: the tip of the tongue taking a trip of three steps down the palate to tap, at three, on the teeth. Lo. Lee. Ta.
>
> She was Lo, plain Lo, in the morning, standing four feet ten in one sock. She was Lola in slacks. She was Dolly at school. She was Dolores on the dotted line. But in my arms she was always Lolita.

He is perhaps one of the greatest novelists of the twentieth century. As an article in the literary supplement of *The New York Times* recently reminded us, 'in academic circles Nabokov is increasingly mentioned alongside names such as Proust and Joyce'.

And yet Nabokov sought, by his own account, a very different kind of renown. One of his poems, 'On Discovering a Butterfly', begins like this: 'I found it and named it, being versed/ in taxonomic Latin; thus became/ godfather to an insect and its first/ describer – and I want no other fame.' Butterflies were his passion. *Lolita* was written during one of the trips he made west every year in the United States, avidly searching for butterflies.

In that serene pantheon where the souls of great writers dwell, I can imagine Nabokov smiling: a few years ago in the *Proceedings of the Royal Society of London*, one of the most authoritative scientific journals, an article was published announcing that his most audacious scientific theory had been confirmed. His name will remain for ever in the annals of science: he was the first to understand the migration of the Blue Icarus (*Polyommatus Icarus*), the enchanting blue butterfly that can be admired in the museum in Milan. *This* was the kind of fame he was looking for: to be 'the godfather of an insect'.

Nabokov's theory was about the mode of migration of these butterflies on the continent of America. In 1945 he published the hypothesis that they had evolved in Asia and had arrived in the United States by crossing the Bering Strait in five successive waves, during the course of 10 million years. No one took him seriously. It was difficult to imagine that butterflies living in warm climates could push so far north. And yet Nabokov was right: modern DNA sequencing techniques have made it possible to reconstruct the genealogy of the species and to confirm his hypothesis exactly. In addition, the reconstruction of changes in climate over time has shown that the Bering Strait underwent phases of sufficiently warm climate to make it possible for the passage of such waves of butterflies, precisely in the periods that Nabokov had suggested.

Nabokov was the curator of the lepidoptera section in the

Harvard University Museum of Comparative Zoology. He published detailed descriptions of hundreds of species. He used to collect butterflies in his childhood, the happy descendant of an extremely wealthy family of the Russian aristocracy. When he was eight years old, his father was imprisoned for political reasons: the young Vladimir carried a butterfly to his cell. With his father murdered and the family fortune lost in the Revolution, he escaped to Europe, where he eventually used the earnings from his second novel to pay for a butterfly-hunting expedition in the Pyrenees.

He was forced to flee from Europe too, after the Nazis came to power, and continued to cultivate his passion for entomology in the United States. He was regarded as a skilful amateur, capable of describing the different species of butterfly, being himself one of the last specimens of a type nearing extinction: nineteenth-century aristocrats who collected lepidoptera as a pastime. But a decade after his death in 1977, various entomologists began to take his scientific work seriously. His classifications turn out to be astute. One of the butterflies he described is named *Nabokovia cuzquenha* in his honour. A book published in 1999, *Nabokov's Blues*, tells the story of the rediscovery of Nabokov's classifications. But another ten years elapsed before the spectacular proof arrived of his hypothesis about butterflies crossing the Bering Strait, and with it the recognition of his status as a scientist of real worth.

Is there a connection between Nabokov's science and his literary work? It is hard to resist the temptation of associating Lolita with butterflies, especially the Lolita seen through the lens of Humbert Humbert's desperate love. But this is probably too facile. The issue is discussed in an essay by Stephen Jay Gould with the suggestive title 'There is No Science

9

without Imagination, and No Art without Facts: The Butter-flies of Vladimir Nabokov', in which he argues that Nabokov's acute focus, his almost obsessive concern for observation and detail, is at the root of both his success as a butterfly collector and his technique as a novelist. Which is probably true. Nabokov himself has written: 'A writer must have the pre-cision of a poet and the imagination of a scientist.'

To me this doesn't seem enough. In 1948, in a passage inserted into *Speak, Memory*, one of the most celebrated liter-ary biographies of the twentieth century, Nabokov writes in his luxuriant, exacting prose:

> The mysteries of mimicry had a special attraction for me. Its phenomena showed an artistic perfection usually associated with man-wrought things. Consider the imitation of oozing poison by bubblelike macules on a wing (complete with pseudo-refraction) or by glossy yellow knobs on a chrysalis ('Don't eat me – I have already been squashed, sampled and rejected'). Consider the tricks of an acrobatic caterpillar (of the Lobster Moth) which in infancy looks like bird dung, but after molting develops scrabbly hymenopteroid append-ages and baroque characteristics, allowing the extraordinary fellow to play two parts at once (like the actor in Oriental shows who *becomes* a pair of intertwined wrestlers): that of a writhing larva and that of a big ant seemingly harrowing it. When a certain moth resembles a certain wasp in shape and color, it also walks and moves its antennae in a waspish, unmothlike manner. When a butterfly has to look like a leaf, not only are all the details of a leaf beautifully rendered but markings mimicking grub-bored holes are generously thrown in. 'Natural selection', in the Darwinian sense, could not explain the miraculous coincidence of imitative aspect

and imitative behaviour, nor could one appeal to the theory of 'the struggle for life' when a protective device was carried to a point of mimetic subtlety, exuberance and luxury far in excess of a predator's power of appreciation. I discovered in nature the nonutilitarian delights that I sought in art. Both were a form of magic, both were a game of intricate enchantment and deception.

There's a lot more here than the capacity to notice details with obsessive attention. There is also, not least, the capacity to see beauty.

Even when our attention alights on something momentarily and then slides away. On the wings of a butterfly. Or the sound – 'Lo-li-ta' – of an unforgettable name.

Newton the Alchemist

In 1936 Sotheby's puts up for auction a collection of unpublished writings by Sir Isaac Newton. The price is low, £9,000; not much when compared to the £140,000 raised that season from the sale of a Rubens and a Rembrandt. Among the buyers is John Maynard Keynes, the famous economist, who was a great admirer of Newton. Keynes soon realizes that a substantial part of the manuscript writings deal with a subject that few would have expected Newton to be interested in. Namely: alchemy. He sets out to acquire all of Newton's unpublished writings on the subject, and soon realizes further that alchemy was not something that the great scientist was marginally or briefly curious about: his interest in it lasted throughout his life. 'Newton was not the first of the Age of Reason,' he concludes, 'he was the last of the magicians.'

In 1946 Keynes donated his unpublished Newtoniana to the University of Cambridge. The strangeness of Newton in alchemical guise, seemingly so at odds with the traditional image of him as the father of science, has caused the majority of historians to give the subject a wide berth. Only recently has interest in his passion for alchemy grown. Today a substantial amount of Newton's alchemical texts have been put online by researchers at the University of Indiana and are now accessible to everyone.* Their existence still has the capacity to provoke discussion, and to cast a confusing light over his legacy.

* http://webapp1.dlib.indiana.edu/newton/.

Newton is central to modern science. He occupies this pre-eminent place because of his exceptional scientific results: mechanics, the theory of universal gravity, optics, the discovery that white light is a mixture of colours, differential calculus. Even today, engineers, physicists, astronomers and chemists work with equations written by him, and use concepts that he first introduced. But even more important than all this, Newton was the founder of the very method of seeking knowledge that today we call modern science. He built upon the work and ideas of others: Descartes, Galileo, Kepler, etc., extending a tradition that goes back to antiquity; but it is in his books that what we now call the 'scientific method' found its modern form, immediately producing a mass of exceptional results. It is no exaggeration to think of Newton as the father of modern science. So what on earth does alchemy have to do with any of this?

There are those who have seen in these anomalous alchemical activities evidence of mental infirmity brought on by premature ageing. There are others who have served their own ends by attempting to enlist the great Englishman among critics of the limitations of scientific rationality.

I think things are much simpler than this.

The key lies in the fact that Newton never published anything on alchemy. The papers that show his interest in the subject are extensive, but they are all unpublished. This lack of publication has been interpreted as a consequence of the fact that alchemy had been illegal in England since as early as the fourteenth century. But the law prohibiting alchemy was lifted in 1689. And besides, if Newton had been so worried about going against laws and conventions, he would not have been Newton. There are those who have portrayed him as some kind of demonic figure attempting to glean extraordinary and ultimate knowledge that he wanted to keep exclusively for himself, to enhance his own power. But Newton really had made extraordinary discoveries, and had not sought to keep those to himself: he published them in his great books, including the *Principia*, with the equations of mechanics still used today by engineers to build aeroplanes and edifices. Newton was renowned and extremely well respected during his adult life; he was President of the Royal Society, the world's leading scientific body. The intellectual world was hungry for his results. Why did he not publish anything based on all those alchemical activities?

The answer is very simple, and I believe that it dispels the whole enigma: he never published anything because he never arrived at any results that he found convincing. Today it is easy to rely on the well-digested historical judgement that alchemy had theoretical and empirical foundations that were far too weak. It wasn't quite so easy to reach this conclusion in the seventeenth century. Alchemy was widely practised and studied by many, and Newton genuinely tried to understand whether it contained a valid form of knowledge. If he had found in alchemy something that could have withstood the method of rational and empirical investigation that he himself was promoting, there can be no doubt that Newton would have published his results. If he had succeeded in extracting from the disorganized morass of the alchemical world something that could have become science, then we would surely have inherited a book by Newton on the subject, just as we have books by him on optics, mechanics and universal gravity. He did not manage to do this, and so he published nothing.

Was it a vain hope in the first place? Was it a project that should have been discarded even before it began? On the

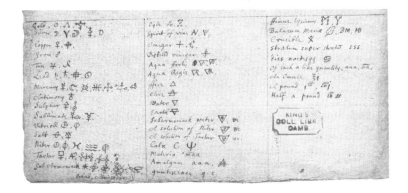

A manuscript page of Newton's, containing a list of alchemical symbols

contrary: many of the key problems posed by alchemy, and quite a few of the methods it developed, in particular with reference to the transformation of one chemical substance into another, are precisely the problems that will soon give rise to the new discipline of chemistry. Newton does not manage to take the critical step between alchemy and chemistry. That would be down to scientists of the next generation, such as Lavoisier, to achieve.

The texts put online by the University of Indiana show this clearly. It is true that the language used is typically alchemical: metaphors and allusions, veiled phrases and strange symbols. But many of the procedures described are nothing more than simple chemical processes. For example, he describes the production of 'oil of vitriol' (sulphuric acid), *aqua fortis* (nitric acid) and 'spirit of salt' (hydrochloric acid). By following Newton's instructions, it is possible to synthesize these substances. The very name that Newton used to refer to his attempts at doing so is a suggestive one: 'chymistry'. Late, post-Renaissance alchemy strongly insisted on the experimental verification of ideas. It was already beginning to face in the direction of modern chemistry. Newton understands that somewhere within the confused miasma of alchemical recipes there is a modern science (in the 'Newtonian' sense) hidden, and he tries to encourage its emergence. He spends a great deal of time immersed in it, but he doesn't succeed in finding the thread that will untie the bundle, and so publishes nothing.

Alchemy was not Newton's only strange pursuit and passion. There is another one that emerges from his papers that is perhaps even more intriguing: Newton put enormous effort into reconstructing Biblical chronology, attempting to assign precise dates to events written about in the holy book. Once again, from the evidence of his papers, the results were not great: the father of science estimates that the beginning of the world happened just a few thousand years ago. Why did Newton lose himself in this pursuit?

History is an ancient subject. Born in Miletus with Hecateus, it is already fully grown with Herodotus and Thucydides. There is a continuity between the work of historians of today and those of antiquity: principally in that critical spirit that is necessary when gathering and evaluating the traces of the past. (The book of Hecateus begins thus: 'I write things that seem to me to be true, because the tales told by the Greeks seem to me full of contradictory and ridiculous things.') But contemporary historiography has a quantitative aspect linked to the crucial effort to establish the precise dates of past events. Furthermore, the critical work of a modern historian must take into account all the sources, evaluating their reliability and weighing the relevance of information furnished. The most plausible reconstruction emerges from this practice of evaluation and of weighted integration of the sources. Well, this quantitative way of writing history begins with

Newton's work on Biblical chronology. In this case too, Newton is on the track of something profoundly modern: to find a method for the rational reconstruction of the dating of ancient history based on the multiple, incomplete and variably reliable sources that we have at our disposal. Newton is the first to introduce concepts and methods that will later become important, but he did not arrive at results that were sufficiently satisfactory, and once again he publishes nothing on the subject.

In both cases we are not dealing with something that should cause us to deviate from our traditional view of the rationalistic Newton. On the contrary, the great scientist is struggling with real scientific problems. There is no trace of a Newton who would confuse good science with magic, or with untested tradition or authority. The reverse is true; he is the prescient modern scientist who confronts new areas of scientific enquiry clear-sightedly, publishing when he succeeds in arriving at clear and important results, and not publishing when he does not arrive at such results. He was brilliant, the most brilliant – but he also had his limits, like everyone else.

I think that the genius of Newton lay precisely in being aware of these limits: the limits of what he did *not* know. And this is the basis of the science that he helped to give birth to.

Copernicus and Bologna

On 6 January 1497, paying nine *grossetti* for the privilege, a young Pole registered for an academic place at the University of Bologna, signing himself 'Dominus Nicolaus Kopperlingk de Thorn'. After six years spent studying in Italy – in Bologna, Padua, Rome and Ferrara – Copernicus (as he came to be known) returned to Poland and devoted the rest of his life to developing a new model of the universe. He would become there the author of a book explaining that new conception – *On the Revolutions of the Heavenly Spheres* (*De revolutionibus orbium coelestium*) – one of the most important works in the history of humanity. Thanks to his book, this species of little creatures living on a marginal planet, of a peripheral star, in one of the billions of galaxies in the cosmos, realizes for the first time, with utter astonishment, that they are not the centre of the universe.

What role did the years spent at an Italian university have, in preparing Copernicus to make this fundamental step-change to our civilization?

I think that the answer is twofold. Copernicus discovered two treasures in Italy. First, he discovered the books that contained, as in a casket, the knowledge accumulated by humankind. He found Ptolemy's *Almagest* and the *Elements* of Euclid, works that summarized the best of the great astronomical and mathematical knowledge of antiquity. He found Italian astronomers such as Domenico Maria Novara, with whom he became very close, who knew how to understand such texts and introduced them to him. He learned Greek and had access to the texts where he probably encountered the heliocentric ideas of Aristarchus; and to the Arabic manuscripts where he was able to study those attempts to retouch the Ptolemaic astronomical system that had been made for a millennium.

But this rich cultural legacy had been available for many centuries. It was available to Indian, Persian, Arabic and Byzantine astronomers, who all had recourse to it. Yet none of them knew how to use such a legacy to understand the crux of the matter: that we do not live at the centre of the universe. Copernicus must have had something else, something extra available to him that enabled this great leap. What was it?

The years that Copernicus spends in Italy include those in which the twenty-three-year-old Michelangelo sculpts his *Pietà* and Leonardo da Vinci tests his flying machines and paints his *Last Supper*. The new, luminous cultural fervour of that Italian humanism which ushers in the Renaissance was stirring in the Italian universities, and in courts such as that of Lorenzo de Medici, where voices were sounding that would have been unthinkable just a short time ago: 'How beautiful youth is, how nonetheless fleeting! Let anyone seeking happiness enjoy it: tomorrow brings nothing but uncertainty . . .' Research into ancient texts and the rediscovery of the knowledge of the past – the obsession of humanists – was being propelled by a burning desire to innovate a new future entirely different from the present.

Petrarch had begun the previous century by writing that: 'The works of the past are like the flowers from which bees collect nectar to make honey.' And the honey was really starting to flow in Italy at the turn of the fifteenth and sixteenth centuries. The spirit of the time constituted a profound opening on to the radically new, as can be seen in the art of the period. It was nothing less than a faith in an alternative world, far different from the structured and hierarchical mental universe of the Middle Ages. Intellectual freedom, courage in pursuing and upholding individual ideas, rebellion against the grand, rigid systems of medieval thought: this spirit of innovation, this deep-seated revolt against given circumstances is the second great intellectual resource that Copernicus was about to partake of when he paid his nine *grossetti* to enroll at the University of Bologna. He doesn't just find Euclid, Ptolemy and Aristotle in Italy: he also finds the idea that their great knowledge can be revolutionized.

I believe that this two-fold experience is what a great university can offer to all of us.

For me, my time at Bologna involved the discovery of extraordinary ideas and texts, such as the works of Einstein, or Paul Dirac's seminal book *The Principles of Quantum Mechanics*, the fundamental work on the subject. I came across the latter because my professor of applied mathematics, Guido Fano, assigned to me a study of the application of group theory to quantum mechanics, an area of physics that I knew nothing about. And so I began to research it – experiencing a fascination with the subject that would continue for the rest of my life. This intellectual richness, discovered in Bologna, turned out to be utterly crucial for me.

But I also found something else in Bologna, when I studied there in the seventies: an encounter with that spirit of my generation, a generation that was intent on changing everything, that dreamed of inventing new ways of thinking, of living together and of loving. The university was occupied for several months by politically engaged students. I got involved with the friends of Radio Alice, the independent radio station that had become the voice of the student revolt. In the houses we were sharing, we nourished the adolescent dream of starting from zero, of remaking the world from scratch, of reshaping it into something different and more just. A naïve enough dream, no doubt, always destined to encounter the inertia of the quotidian; always likely to suffer great disappointment. But it was the same dream that Copernicus had encountered in Italy at the beginning of the Renaissance. The dream not only of Leonardo and of Einstein but also of Robespierre, Gandhi and Washington: absolute dreams that often catapult us against a wall, that are frequently misdirected – but without which we would have none of what is best in our world today.

What can the university offer us now? It can offer the same riches that Copernicus found: the accumulated knowledge of the past, together with the liberating idea that knowledge can be transformed and become transformative. This, I believe, is the true significance of a university. It is the treasure-house in which human knowledge is devotedly protected, it provides the lifeblood on which everything that we know in the world depends, and everything that we want to do. But it is also the place where dreams are nurtured: where we have the youthful courage to question that very knowledge, in order to go forward, in order to change the world.

My 1977, and That of My Friends

I can't identify with most of the articles I have read recently about the youth movement that swept through Italy forty years ago now, in 1977, as briefly and intensely as a storm. It seems to me that they are not referring to what was actually said, thought and felt by my friends and I, all those years ago. I'm not about to attempt a historical or sociological analysis, and I have no desire to mistake my own experience and that of some of my friends for some kind of historical fact. But at the time I know that there were many who felt as I did, and somewhere they must still exist. I'm writing this with them in mind, the many friends I had then, as well as for those who are curious about hearing an alternative version of the period.

Some of those friends have maintained a somewhat rose-tinted view of what, in their eyes, has become an almost mythical time. It was a moment of intense dialogue, of dreams, enthusiasm, yearning for change, of longing to build together an alternative and better world, which they now remember with equally intense nostalgia. So much so that they have tended to present everything that has happened since in our lives as drab by comparison. This is definitely not how I feel. We were in our twenties, and life at that age is frequently wonderful, and our experience of it heightened – especially in our memory of it. This is not the fragrance of history, it is the fragrance of youth. For me those events remain remarkable, even magical, but precisely due to the fact that they were the beginning of something. A path opened up for me. The life that followed was not greyer: I was part of a collective discovery of a spectrum of colour, and those colours have stayed with me. That said, the year after 1977 was undoubtedly experienced by many of us as a defeat. The bright desire to change the world which had momentarily seemed to open on to genuine possibilities had collided with a harsh reality. Shipwrecked by the violence of the reaction of the State, which we called at the time the repression, and then by the violence that we now call terrorism. There were many of us who believed and said that nothing good would come of the 'armed struggle' in Italy, that it was merely an extreme and foolish reaction, a desperate result of dreams that were over. The 'wayward comrades', many of us knew, were young men and women with a more absolute sense of morality than the rest of us; and therefore, as is unfortunately often the case, blinded by it. We wanted something different, and for a brief moment, along with many others, we had thought that change was coming; that it was possible to head in that direction.

Which direction? Dreams have a tendency to seem inconceivable as soon as they are over. But history shows that it is sometimes some of the most apparently inconceivable dreams that become reality: against the expectations of the 'realists', the French Revolution succeeds in bringing down the aristocracy and the *ancien régime*; Christianity prevails over pagan imperial Rome; a pupil of Aristotle conquers the world, and his friends establish libraries and centres of learning and research; the adherents of an Arab preacher transform the thoughts and lives of hundreds of millions of individuals . . . And so on.

More frequently, the big dreams founder against the force of daily life. They are short-lived or even momentary intervals; they come crashing down and are consigned to oblivion. History has so many streams that lead nowhere. The sects of the fourth century that wanted a church that was poor; the egalitarian mirage of Soviet Communism, the recent idea of restoring the Caliphate . . . But what frequently happens is more complex, and history follows tortuously winding courses. The Directory executes Robespierre, Wellington defeats Napoleon, and the King of France is restored to the throne. The revolution is quelled . . . But has it really been lost? Historical movements are made by ideas, ethical judgements, passions, ways of seeing the world. They often lead to a dead end. Sometimes, however, they leave behind traces that continue to work deeply upon the mental fabric of civilization, changing it irreversibly. Revolution is an old mole that burrows deep into the soil of history. On occasion, it pops its head out. It is the fantasy of those who rule that nothing will change. But then, the old mole appears when least expected. Our civilization, the set of values in which we believe, is the result of countless ideals, of the vision of those who have dared to look and to dream, intensely, beyond the present.

What is known as the Movement of 1977 in Italy is unintelligible when regarded in isolation. It was a late expression – not the last but one of the last, self-consciously so, and as a result of this all the more intense – of that dream which swept over not just Italy but the whole world in the sixties and seventies. These were years in which a significant global youth movement dared to dream, and to fervently hope that radical social change was possible. It certainly wasn't a movement with a structured and coherent set of aims; in reality, it consisted of a thousand different streams. But despite their great diversity, all of these streams felt that they were part of the same river, sharing the same current – from the squares of Prague to the universities of Mexico City; from the campus of Berkeley to Piazza Verdi in Bologna; from hippy communes in California, both rural and urban, to South American guerrillas. And from Catholic marches in support of the Third World, to English experiments in anti-psychiatry, and from Taizé to Johannesburg. Despite the huge differences in specific attitudes, there was a prevailing awareness of belonging to the same great flow, of sharing a single great dream. Of being part of the same 'struggle', as it was commonly referred to at the time, to bring into existence a very different world.

It was a dream of building a world where there would be no huge social inequalities, no male domination of women: a place without borders, without armies, without poverty. It was the idea of replacing all competitive struggle for power with cooperation, of leaving behind the bigotries, fascisms, nationalisms, the narrow identitarianism that had led the preceding generations to exterminate 100 million human beings in two world wars. These were far-reaching dreams, envisaging a world without private property, without envy or jealousy, without hierarchy, without churches, without powerful states, without atomistic closed family units, without dogma. A world, in a word, that was free. Somewhere with no need for the excesses of consumerism, where you could work for pleasure, not for getting, spending or social climbing.

Just mentioning these ideas today could lead some to say you are delirious. And yet there were plenty of us then who believed in them, all over the world. At that time, I travelled a great deal, on several continents, and everywhere I went I would come across young people with the same dreams. This is what my friends and I were talking about in 1977. Not, certainly, about today's worries, such as financial insecurity or the fear of not finding a job. We did not want a job: we wanted to be free from labour. If we want to recall something of those years, this is what I remember most vividly.

We lived in open houses. We slept here and there. We knew how dangerous heroin was, and anyone with an ounce of sense avoided it. But we also believed that marijuana and LSD were not harmful, and joints would be passed around as naturally as today we would offer and accept a glass of wine. LSD was something else: a transformative and significant experience, to be treated with caution and respect – but one from which we could learn a great deal. Our principal occupation, as with every youth, was to fall madly in love and to lose ourselves in passions. Sex was a kind of common currency, a way of encountering and getting to know others – men and women alike. It was taken seriously, regarded as the centre of life, almost like a religion. And just as every religion fills the lives of believers, we wanted to fill our lives with love and with lovemaking. As well as with friendship, music and the invention of new ways of being together that were different from the grey, competitive ones of other generations. We were experimenting with communal living, with the exclusion of jealousy, trying to really coexist. Of course there were arguments and fallings-out, as in any family; but the feeling of being part of a huge family that was scattered throughout the world was a deep-seated one: an enormous family that was moving together, like interplanetary explorers aspiring to create a new and different world. I have always thought that the Quakers in the first European communities in America, the apostles of Jesus in Palestine, the first Christians, the young

26

Italians of the Risorgimento, the companions of Che Guevara in Bolivia, even Plato's pupils in the Academy must have felt a little like we felt . . .

Yet we failed completely to build a new world. We were soon enough disillusioned. Some plans were abandoned because they were mistaken; many others because they were defeated. The plausibility of our aspirations melted like snow in sunlight. We went our separate ways, each one following their own path.

Was it futile to have dreamed at all? I don't think so, for two reasons. The first is that for many of us those dreams fertilized the ground from which our lives grew. Some values from that time remained deeply rooted in us – and we were carried forward by those aspirations. The extreme form of freethinking cultivated in those years, whereby everything seemed possible and worth exploring, and every idea could be modified and adapted, provided for many of us the source for whatever it was that we went on to do with our lives.

I don't know if the second reason is believable or not. But it exists all the same. Dreams of building a better world have frequently been defeated in the course of history, but they went underground and continued to be active there. And in the end, they contributed to real change. I still believe that this world, which seems ever more filled with war, violence, extreme social injustice and bigotry, with nationalistic, racial and regional groups attempting to wall themselves into their own narrow identities and to fight against each other, is not the only world possible. And in this, perhaps I am not alone.

Literature and Science:
A Continuing Dialogue

The greatness of literature lies in its capacity to communicate the experiences and feelings of human beings in all their variety, affording us glimpses of the boundless vastness of humanity. Literature has told us about war, adventure, love, the monotony of everyday life, political intrigues, the life of different social classes, murderers, banal individuals, artists, ecstasy, the mysterious allure of the world . . . Can it also tell us anything about the real and profound emotions connected with great science?

Of course it can: literature is full of science. An entire genre, science fiction, is fed by it. Playwrights have engaged with it, not the least of them Brecht in his *Galileo*, a play that goes to the heart of the critical attitude upon which scientific thought is based:

> We put everything, everything in doubt [. . .] What we find today, tomorrow we will erase from the blackboard and we will not write any more, at least until we find it again the day after that. If some discoveries follow our predictions, we will look on them with particular distrust. [. . .] And only when we have failed, when beaten and without hope we are reduced to licking our wounds, then with iron in our souls we will begin to ask ourselves if we might not be right after all.

This is how, at the end of the play, Brecht has Galileo respond to his young assistant Andrea, when the latter is impatient to immediately find evidence to corroborate a brilliant idea. Many great scientists, and Galileo himself with his *Dialogue Concerning the Two Chief World Systems*, have written works that indisputably figure as classics of literature as well as of science.

But it is the greatest literature that has sought to come to terms directly with the scientific vision of the world. The striking opening of one of the most intelligent novels of the early twentieth century, Robert Musil's *A Man without Qualities*, is a dry list of meteorological data that opens out, at the end of the paragraph, with their translation into vernacular and everyday terms: '. . . it was, in other words, a beautiful August day'. Here, and filigreed throughout the novel, Musil attempts to incorporate and come to terms with a vision of the world revealed by the great successes of nineteenth-century science: a world of data and numbers.

The same challenge, if very differently inflected, was faced by the Milton of *Paradise Lost*. The poem includes these tremendous lines in which he wonders about the Copernican model that was still hypothetical at the time:

What if the Sun
Be Centre to the World, and other Stars
By his attractive virtue and their own
Incited, dance about him various rounds?
Their wandering course now high, now low, then hid,
Progressive, retrograde, or standing still,
In six thou seest, and what if sev'nth to these
The Planet Earth, so stedfast though she seem,
Insensibly three different Motions move?

The passage brims with the excitement of an immense step forward in science, a radical remapping of the universe that was in the process of being accomplished. All of Milton is secretly oxygenated by the new science: the immensity of the cosmos, the harmonious yet complex nature of the universe and of its movements, interstellar space and the possibility of travelling through it, the dominant role of the sun, the probability of extraterrestrial life ... Throughout Milton's writings, there is the impetus of the great conceptual revolution that in the seventeenth century was being brought about by science.

But to find a pure singer of science we need to go back further, until we get to the great poet who was able to devise a way of thoroughly uniting poetry and science, demonstrating how intimately linked they really are, to the point of almost becoming the same thing. I am talking about Lucretius, in whose work the most rational of deductions acquires the power of poetry:

And now if we accept that the number of atoms is so endless
That an entire human era would not be sufficient to count them,
And that if there exists the same force and nature that may
Bring these atoms together anywhere, in the same fashion
That they have converged here, then it is necessary to acknowledge
That there must be other terrestrial globes elsewhere in the void
And different races of men, and different species of beasts.

Naturalism, which animates science, was not only the source of Leopardi's anguish, but filled Lucretius with a kind of serenity: 'Sometimes, like children who are afraid of the dark, we fear in the light of day things as inconsistent as those that the child is afraid of at night.' And it is this thoroughgoing naturalism that allows Lucretius, the anti-religious classical writer par excellence ('How many afflictions have been brought about by religion . . .'), to turn to the goddess Venus with such luminous sentiments:

Mother of Aeneas and all his race, delight of men and of gods,
Alma Venus, who beneath the wandering stars of the heavens
Populates with living creatures the sea furrowed by ships, the earth
Fecund with fruits; through you every living species forms,
And once it has blossomed can come out to see the light:
Before you, O Goddess, the winds run – at your first appearance
The clouds leave the sky – for you the ever industrious earth
Brings forth sweet-smelling flowers; for you expanses of the seas
laugh, and the becalmed sky is radiant with light.

Twenty centuries have elapsed since Lucretius, during which abysses of new knowledge – and alongside it new, boundless mysteries – have gradually opened up before us. Will we be able to find someone capable today of singing, with as much lucidity, about the complexity and mystery, as well as the strange comprehensibility and profound beauty of nature, as revealed by the lights of science?

Dante, Einstein and the Three-Sphere

Having climbed to the outermost sphere of the Aristotelian universe, Dante is invited by Beatrice to look down. He sees the entire heavens, and down there, at the bottom, the Earth, which seems to be slowly revolving beneath his feet. But then Beatrice invites him to look up, beyond the Aristotelian universe, where according to Aristotle there would be nothing, since for Aristotle the universe has a definite border where everything ends. Dante looks, and has an extraordinary vision of a point of light surrounded by nine immense spheres of angels. Where is this point of light, and where are the angelic spheres located, since they are outside the great globe of the universe, as conceived by Aristotle? Dante says intriguingly that this other part of the universe 'surrounds the first in a circle, like the first surrounds the others', and in the next canto refers to it as 'appearing to be enclosed by that which it encloses'. The point of light and the spheres of angels surround the universe, and at the same time are surrounded by it.

What does this mean? For the majority of readers, the notion of two sets of concentric spheres, each of which 'encloses' the other, is just an obscure poetic image. Senior-school textbooks in Italy simply draw the point of light and the spheres outside of the Aristotelian universe. But for a contemporary mathematician or cosmologist, the description of the universe is perfectly clear, and the object described by Dante is unmistakable. He is describing a 'three-sphere', the shape that in 1917 Albert Einstein hypothesized was the shape of our universe, and that today remains compatible with the most recent astronomical measurements. Dante's unbridled poetic imagination and extraordinary intelligence anticipated by centuries a brilliant intuition of Einstein's on the shape that our universe might have.

What is a 'three-sphere'? It is a mathematical structure, a geometrical figure that is not so easy – but ultimately not so difficult either – to conceive. In order to understand it, think about the following problem: if I keep walking in the same direction on Earth, where will I eventually end up? Will I come to the edge of the Earth? No. Will I carry on reaching an infinite number of new countries? Not at all. As everyone knows, once we have gone around the Earth, we find ourselves back where we started. This was a difficult idea for our ancestors to digest, and it still makes primary schoolchildren laugh, but it is one that we became habituated to in the end, and that we now find perfectly reasonable. This is because the Earth is a 'sphere'. Mathematicians, more precisely, say that the 'topology' of the surface of the Earth, which is to say its 'intrinsic form', is a 'two-sphere' ('two' because on Earth we can walk in two main directions: north–south or east–west). Let's ask the same question for the entire universe in which we find ourselves. Let's imagine that we can travel in an immensely fast spaceship, always in the same direction. Where would we end up? Would we reach the edge of the universe? This is highly unlikely. Would we go on discovering infinite amounts of new space? This is not a very appealing idea either, and is scarcely credible. So what next? There is a third possibility: after having journeyed for long enough we

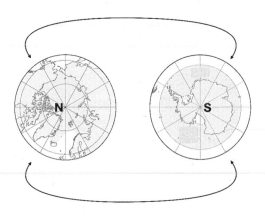

will arrive back where we started from, on Earth, having 'gone around' the universe. This is what happens if the universe is a three-sphere.

The difficulty of actually visualizing a three-sphere lies in the fact that it can't be contained within the space that we are used to. It's the same reason that the surface of the Earth cannot be decently represented in a flat geographical map. That said, there is a simple way of conceiving a three-sphere. Think again about the surface of the Earth. One way of reproducing it on a geographical map consists of drawing two discs: one with the continents belonging to the northern hemisphere, with the North Pole at the centre, and the other using the same method for the southern half of the globe. The equator is drawn twice, in the form of the edge of both discs. If we start out from the South Pole and head north, at a certain point we will cross the equator: in our representation of the surface of our planet in two discs, we 'leap' from one disc to the next. Obviously, in reality, no such leap occurs, because the northern hemisphere seen by someone coming from the south 'surrounds' the southern hemisphere,

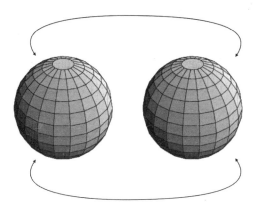

just as the southern hemisphere 'surrounds' the northern hemisphere for anyone looking from the north.

Well, the three-sphere can be represented in the same way, by using two balls. One ball is the 'southern hemisphere' of the three-sphere, the other is the 'northern hemisphere'. The 'equatorial' sphere that separates and connects the two hemispheres is represented twice: it is the border between the two balls.

A traveller starting from the centre of the first ball and ascending, like Dante, 'from sphere to sphere', when he reached the equator (the surface of the balls) would see beneath him a combination of concentric spheres, and above him another identical set of concentric spheres that would be enclosed around a point. This other hemisphere would at the same time 'surround' and 'be surrounded by' the first ball. In other words, this representation of the three-sphere corresponds to that of Dante.

The American mathematician Mark Patterson was the first to write, in an article published in 1979, about how Dante had very clearly described the three-sphere. But today any physicist or mathematician would easily recognize the three-sphere in Dante's description of the universe.

How was Dante able to anticipate Einstein by six centuries? I think that one reason may be the fact that Dante's spatial imagination, so to speak, was medieval and therefore not yet confined by the rigid Newtonian version of physics according to which physical space is Euclidean and infinite. For Dante, just as for Aristotle, space is only the structure of the relations between things, and that structure may adopt peculiar shapes. Furthermore, the idea that the divine resides 'beyond' the border of the universe can be found in the *Tresor*, a wonderful compendium of medieval knowledge written by Dante's mentor, Bruno Latini. The idea of God as a point of light surrounded by angels was already present in the period and can be found in various medieval images. Dante put together in an original and intelligent way the pieces of a puzzle that already existed.

I believe that Dante may have been inspired by one image in particular. Dante left Florence in 1301, while the extraordinary mosaics in the cupola of the Baptistery were being completed. If you were to visit the Baptistery today and look up, you would see a point of light (natural light coming through the lantern window at the apex of the cupola) surrounded by nine orders of angels, with the name of each order clearly labelled: Angels, Archangels, Principalities, Powers, Virtues, Domains, Thrones, Cherubim and Seraphim – exactly as they appear in the *Paradiso*. Imagine that you are an ant on the floor of the Baptistery (at the South Pole) and that you begin to crawl in any direction; regardless of the point at which you then begin to scale the wall, you will inevitably arrive at the point of light surrounded by angels (the North Pole). The point of light and its angels 'surrounds' and at the same time 'are surrounded by' the rest of the decorated interior of the Baptistery. That interior is clearly a two-sphere. Dante, like every other citizen of Florence at the end of the thirteenth century, would undoubtedly have been impressed by the grand architectural project that his city was completing. The magnificent and terrifying mosaic depicting Hell, the work of Coppo di Marcovaldo, the teacher of Cimabue, is widely regarded as a source of inspiration for Dante's own *Inferno*. Is it not possible that he also found inspiration for the very shape of his universe from the internal structure of the Baptistery? The *Paradiso* reproduces its structure, with its rings of angels and its point of light, translated from two dimensions into three – and in doing so arrives at Einstein's three-sphere.

Whether it was this or something else that gave him the idea, the fact remains that Dante's extraordinary imagination was able to find a solution to the ancient problem of how to reconcile a finite world with one that also has no 'edge'. And this solution is, amazingly, the same one that with mathematical exactitude Einstein will arrive at six centuries later. What's more, it might well turn out to be the right solution.

Why are we so keen on Dante? For many reasons, including perhaps one that as a scientist I can readily appreciate: Dante was not only highly cultivated but also possessed an extraordinary intelligence that encompassed scientific understanding. To hear a cultivated person of today joking almost boastfully that they are completely ignorant about science is as depressing as hearing a scientist bragging that they have never read a poem.

Poetry and science are both manifestations of the spirit that creates new ways of thinking the world, in order to understand it better. Great science and great poetry are both visionary, and sometimes may arrive at the same insights. The culture of today that keeps science and poetry so far apart is essentially foolish, to my way of thinking, because it makes us less able to see the complexity and the beauty of the world as revealed by both.

Between Certainty and Uncertainty: A Precious Intermediate Space

In the institute where I used to work a few years ago, a rare non-infectious illness hit five colleagues, one after the other, in short succession. There was a sense of alarm, and a hunt for the cause of the problem. At first, we thought that there might be some sort of chemical contamination within the buildings of the institute, since they had been used in the past as a biology lab. But nothing of this kind was found. The general level of apprehension grew, and some even looked for work elsewhere.

One evening at a dinner party, I mentioned these events to a friend of mine who is a mathematician, and he burst out laughing. 'There are four hundred tiles on the floor of this room; if I throw one hundred grains of rice into the air, will I find,' he asked us, 'five grains on any one tile?' We replied in the negative: there would be one grain for every four tiles. We were wrong. We actually tried numerous times, experimentally throwing the rice, and there was always a tile with two, three, four, even five or more grains on it. Why? Why would grains 'flung randomly' not arrange themselves into good order, equidistant from each other? Because they land, precisely, by chance, and there are always disorderly grains that fall on tiles where others have already gathered. Suddenly the strange case of my five colleagues seemed very different. Five grains of rice falling on the same tile does not mean that the tile possesses some kind of 'rice-attracting' force. Five people falling ill at the institute did not mean that it must be contaminated.

Lack of familiarity with statistical thinking is widespread even among the educated, and it is deleterious. The institute where I worked was part of a university. Even know-all professors such as we were had fallen into a gross statistical error. We had become convinced that the 'above average' number of sick people required an explanation. Some had even gone elsewhere, changing jobs for no good reason. Life is full of stories such as this.

It is not at all rare to hear a news programme report as a relevant factor that in a certain locality the percentage of something is above average. The percentage of anything, however, is above average in more or less half of these localities – and below average in the other half. A few years ago, Italians were moved to tears on seeing a television report on cancer patients apparently cured after being treated by an alternative doctor called Di Bella. What better proof of the efficacy of his treatment than to see those with very serious tumours recovering from their illness? Yet this proved to be a foolish assumption. With or without the application of the cure, there are natural forms of recovery from even the gravest tumours. To exhibit recoveries, even when these seem numerous, is not at all the same as proving that the treatment has had a positive effect. To discover whether or not the treatment is effective, we need to count how many times it has worked and how many times it has failed, and compare the results with those of untreated patients. If we don't proceed in this way, we might as well start dancing to make the rain fall from the sky: there will always be days on which our dance will be followed by rain, and we will be able to point to them as evidence of the efficacy of rain dancing . . .

It is this lack of understanding of statistics that leads so many to be impressed by the cures that occur at the famous Sanctuary of Our Lady of Lourdes, to resort to medicines made of sugar and water, or to lose their lives in dangerous games after having witnessed others participating in the same pursuits without coming to any harm.

We would avoid a great deal of foolishness, and society would gain significant advantages, if children were taught the fundamental ideas of probability theory and statistics: in simple form in primary school, and in greater depth in junior and senior. Reasoning of a probabilistic or statistical kind is a potent tool of evaluation and analysis. Not to have it at our disposal leaves us defenceless. To be unclear about such notions as mean, variance, fluctuations and correlations is a bit like not knowing how to do multiplication or division.

The lack of familiarity with statistics leads people to confuse probability with imprecision. On the contrary, probability and statistics are precision tools that allow us to respond in a reliable way to precise questions. Without them, we would not have anything like the efficacy of modern medicine, quantum mechanics, weather forecasts, sociology . . . In fact, we would be without experimental science in its entirety, from chemistry to astronomy. Without statistics we would have very little idea about how atoms, societies and galaxies operate. It was thanks to statistics, to take a couple of random but significant examples, that we were able to understand that smoking is bad for us, and that asbestos kills.

We use probabilistic reasoning every day. Before making decisions, we weigh up the probability that this or that will follow from them. We have a sense of the average price of petrol, and of its variability, which is to say that we are aware how many distributors adhere or do not adhere to the average price. We intuitively anticipate that variables are correlated: the closer the filling-station is to the centre, the more costly the fuel is likely to be. We distinguish between facts that are improbable and less improbable: the probability of being caught up in a train crash is very small, so we tend to take the train; the probability of being run over by a train while using a closed level crossing is also quite small (the majority who take the risk get away with it), but it is sufficiently significant to dissuade us from the attempt. And again, we understand well enough the difference between coincidences that have happened 'by chance' and facts that are determined by 'a reason'. But we use these ideas in an approximate fashion, often making multiple errors. Statistics sharpen and refine them, giving them a precise definition, allowing us to reliably evaluate, for instance, whether a medicine or a building is dangerous or not. It does this by resorting to a quantitative, rigorous application of the notion of probability.

But what is probability? Despite the efficacy of statistics, the nature of probability is the subject of controversy and of lively philosophical debate. A traditional definition is based on 'frequency': if I roll a die many times, one sixth of these times the number 1 will be uppermost; hence I can say that the probability of rolling a 1 is 1/6. However, this definition is quite weak. For instance, we use probability in situations where the 'event' may not be repeated. I think that there is a good prospect, based on probability, that the article I am writing will be published by the editor of the newspaper for which I write; but it makes no sense to send it multiple times, because the second time he would certainly not publish it.

An alternative understanding of probability is as a 'propensity'. A radioactive atom, according to some physicists, has a 'propensity' to decay during the next half an hour, which is evaluated by expressing the probability that this could happen. This is not a very satisfactory interpretation either: it sounds a bit like the 'dormant virtues' of scholasticism mocked by Molière in *Le Malade imaginaire*: the sleeping pill makes us sleep because it has dormant virtues, and the atom decays because it is predisposed to decay.

Clarity on the concept of probability, in my opinion, may be attributed to an outstanding Italian intellectual who has not received the recognition in his own country that he deserves. The philosopher and mathematician Bruno de Finetti (1906–85) introduced in the thirties an idea that proved to be the key to understanding probability: probability does not refer to the system as such (the dice, the newspaper editor, the decaying atom, tomorrow's weather), but to the knowledge that I have about this system. If I claim that the probability of rain tomorrow is 1/3, I'm not saying anything that pertains to clouds that may already be determined by the current situation regarding winds. I am characterizing my degree of knowledge/ignorance of the state of the atmosphere. This is what probability is all about.

We live in a universe where ignorance prevails. We know many things, but there is a great deal more that we don't know. We don't know who we will encounter tomorrow in the street, we don't know the causes of many illnesses, we don't know the ultimate physical laws that govern the universe, we don't know who will win the next election, we don't really know what is good for us and what is bad. We don't know if there will be an earthquake tomorrow. In this essentially uncertain world, it would be foolish to ask for absolute certainty. Whoever boasts of being certain is usually the least reliable.

But this doesn't mean either that we are completely in the dark. Between certainty and complete uncertainty there is a precious intermediate space — and it is in this intermediate space that our lives and our thoughts unfold.

Bruno de Finetti:
Uncertainty is Not the Enemy

What do we know with certainty? A significant branch of the philosophy of science is leaning towards an answer that has its roots in the work of a relatively little-known Italian: Bruno de Finetti.

De Finetti was born in 1906 in Innsbruck of Italian parents. He began his university studies at the polytechnic in Milan, and in 1936 won a competition for a professorship in mathematics, but failed to be appointed due to a law instituted by the Fascist government: preoccupied by a lack of patriotic fertility, it had prohibited the appointment of unmarried professors. He became an academic lecturer only after the war, teaching first in Trieste and later in Rome. At the centre of his interests was the mathematical theory of probability, to which he contributed theorems that are named after him. But his thinking ranged from politics to didactics, and his most original contribution was to the theory of knowledge. His ideas in this field were regarded as revolutionary at the time and have today become a crucial point of reference for science.

To the question 'What do we know about the world with absolute certainty?', the response suggested by de Finetti would be: nothing. This in itself is hardly remarkable: it is the answer given in antiquity by Pyrrhus of Elis, and in modern times in various forms by some of the greatest philosophers, such as David Hume. But de Finetti identifies with acumen the nature of our knowledge, and understands how, notwithstanding the absence of absolute certainty, it can nevertheless develop in a rigorous and credible way and lead to convictions that are justified and, above all, shared.

At the end of the nineteenth century, an era of triumphs for scientific thought and its applications, science appeared to offer a concrete kind of knowledge: Newton and Maxwell had understood the laws that ultimately govern our world. Logical positivism sought to analyse the way in which science obtained truths, on the basis of direct observations of the world. But it soon encountered serious problems: the realization, for example, that all observation is already coloured by theoretical prejudice, and that therefore there is no such thing as 'pure' observation. The revolutions that occurred in physics in the twentieth century have shown that even immensely successful, amply 'confirmed' theories, such as those of Newton and Maxwell, can turn out to be merely approximations. The historical and evolutionary, rather than definitive, character of our scientific knowledge was emphasized by historians of science such as Thomas Kuhn. Overturning the hopes of positivism, the Austrian philosopher Karl Popper exerted an immense influence upon scientists when he maintained that science is not characterized by the fact that its theories have proved to be true, but only by the fact that they can be proved to be false. Theories are true only to the extent that they have not yet been 'falsified'. This means that there is nothing that we can know with certainty.

So, if we lack absolute certainty, what is the value of knowledge? De Finetti's significance lies in the fact that he understood how we can have a shared and reliable knowledge without absolute certainties. His insight was into the subjective character of probability and the probabilistic, but the convergent character of knowledge. The key that makes this possible is a subtle theorem that we owe to an English mathematician of the eighteenth century, Thomas Bayes. Bayes demonstrated two things. Firstly, that every new piece of empirical evidence modifies the probability of beliefs. Secondly, and crucially, how these modifications lead our beliefs to converge, even if they are different to begin with.

The probability of a thesis is an evaluation of the extent to which we expect that the thesis will be true: it is subjective. But this probability changes with every experience. Bayes' theorem tells us how this works. If my belief implied that an event is probable and it does in fact happen, then my belief is strengthened. Otherwise, my belief is weakened. If I maintain that the majority of stars possess planets, my conviction is reinforced by every new star that I see with an accompanying planet. The theorem explains this in quantitative terms. If we allow that real events influence our beliefs in a similar way, the theorem indicates that our beliefs eventually *converge*: they become largely justified by experience. In this way our knowledge, be it scientific or personal, historical or geographical, can be deeply reliable and rationally well founded, without the requirement of absolute certainty.

This is the key to how scientific knowledge works. I can argue that the Earth is probably flat and unlikely to be spherical, and you can think that the opposite is true. But as we gradually notice together that the shadow of the Earth cast on the moon during an eclipse is round; that the further north we go, the higher the polar star is on the horizon; that Magellan went around the world by following the sun and arrived back in Europe, and so on, the probability that the Earth is flat diminishes until it becomes derisory. This way of thinking never requires us to talk about absolute certainty, of definitive conclusions, which would prevent us from ultimately understanding, and from really expanding our knowledge – but it does allow us to converge on sets of convictions with an arbitrarily high degree of credibility. And this is what our knowledge consists of.

Let's turn to de Finetti's own words, beautifully written in a style somewhat redolent of the pre-war period:

> Science, understood as the discoverer of absolute truth, remains therefore, naturally, a source of disillusion for its lack of absolute truths. If the cold marble idol of a perfect, eternal and universal science that we could only seek to better understand falls and shatters, it is there that suddenly alongside it we find a living creature, a science that our thought freely creates. A living entity: flesh of our flesh, the fruit of our torment, our companion in the struggle . . .

In the Anglophone world, it was thanks to the English philosopher Frank Ramsay that the subjective interpretation of probability was taken seriously at the beginning of the twentieth century. It was only much later, in the fifties, that the significance of de Finetti's writings on the subject was recognized. The American philosopher Leonard Savage, who helped to spread de Finetti's reputation in the English-speaking world, relates how he set out to learn Italian so that he could speak with and learn directly from him. Today de Finetti is well known globally, though much less so in Italy. His manuscripts have been collected and archived in Pittsburgh, at one of the world's leading centres of the philosophy of science. Appropriately enough, perhaps, since he embodies a type of de-provincialized Italian intellectual, open to the world and free from the long-dominant Italian intellectual traditions: the shackles of Crocean idealism (which he dubbed *philosofesserie*, combining the Italian words for 'philosophy' and 'hogwash'), as well as from the legacy of Hegelianism. In the great Italian tradition founded by Galileo, de Finetti was able to make technico-mathematical and humanist-philosophical knowledge converge. His original synthesis of classical empiricism (Hume) and pragmatism (Peirce, James), centred on the notion of subjective probability, is becoming increasingly influential, especially in the philosophy of science, where it offers an elegant and convincing solution to the limitations of Popper's thought. De Finetti was very much ahead of his time. His fundamental text, 'Probabilism: A Critical Essay on the Theory of Probability and on the Value of Science', was published in 1931, and only translated into English more than half a century later, in 1989. And his book, *The Invention of Truth*, though written as long ago as 1934, saw the light of day only in 2006, thanks to the invaluable efforts of his daughter Fulvia. No doubt this lag was partly due to the fact that Fascism in its pomp could not countenance doubts about the truth.

In 1968 de Finetti reproved colleagues for their snobbish disregard of students – remarking, in a reversal of the usual order of things, that 'students should always be listened to'. In 1977 I had the pleasure of sharing something with him: we were both accused of subversive association, and of inciting criminal acts. We both went into hiding, and were both shopped to the police. De Finetti responded cleverly and in style: he let it be known to the police that he would consent to being arrested at the entrance to the Accademia dei Lincei, or Lincean Society, Italy's oldest and most prestigious academy of science, of which he was a member. His crime

49

had been to write an article in support of conscientious objection to military service in Italy.

There is a profound lesson to be learned from de Finetti's ideas, one that I believe relates to us all – to our daily and our spiritual lives, and to our lives as citizens: we cannot get rid of uncertainty. We can diminish it, but we cannot make it disappear. Hence we should not experience it as some kind of nightmare. On the contrary, we should be reconciled to it as our lifelong companion. In the end, it is a kind and good one. It is probability that makes life interesting. It is because of probability that we can be touched by the unexpected. It is probability that allows us to remain open to further knowledge. We are limited and mortal, we can learn to accept the limits of our knowledge – but we can still aim to learn and to look for the foundation of this knowledge. It is not certainty. It is reliability.

Does Science Need Philosophy?

I was invited to give a lecture at the London School of Economics on the theme: 'Does Science Need Philosophy?'. It was intended to be the closing talk at the European Congress on the Philosophy of Physics, and it was meant to respond to a recent series of very negative public comments about philosophy, by some very well-known colleagues of mine. Stephen Hawking had written, for instance, that philosophy was dead now that we had science, and a chapter in the most recent book by the Nobel Prize winner Steven Weinberg was entitled 'Against Philosophy'. I accepted the invitation out of solidarity with philosophers, though without really knowing what I was going to say. I started studying, and soon had a marvellous stroke of luck. Like a schoolboy who has been set a difficult piece of homework and stumbles across someone else's perfect answer to copy, I found that the theme had already been excellently developed in a little-known text by a young man undoubtedly more gifted than myself: Aristotle.

In the fourth century BC, the sons of the best families in Athens studied in Plato's Academy. But the Academy was not the only school in the city: others challenged its primacy, and among these rival academies the one run by Isocrates stood out. Between the schools of Plato and Isocrates there was a fierce rivalry, something like that between Oxford and

Cambridge. But the rivalry was not so much regarding quality as method. Education in the Academy was based on the ideas of Plato, who maintained that in everything it was crucial to study the fundamentals. You didn't learn how to be a court judge, to carve a statue or to govern a city – but rather inquired into the nature of justice, of beauty and of the ideal city. Plato had found a term for his method that would go on to enjoy some considerable success – 'philosophy', which originally referred to this way of educating the young and encouraging the development of their knowledge. Isocrates, in his corner, contested this 'philosophical' approach, considering it to be useless and unfruitful. He wrote, for example, that

> Those who study philosophy may be able to actually do something, but they will invariably do it worse than those who participate directly in practical activities. Whoever pays no attention to philosophical discussions and is inducted directly into a practical activity will be much more successful in every case. As far as the arts and sciences are concerned, philosophy is completely useless.

These are more or less the same sentiments expressed by Hawking and by Weinberg in order to criticize philosophy now. But to this criticism a brilliant reply was given by a young student at the Academy. Aristotle – for it was he – couched that reply in the form of a dialogue, Platonic style, and gave it the title *Protrepticus*, which roughly means an 'invitation' (to philosophy). Aristotle responds to the criticisms levelled by Isocrates, and discusses why philosophy, the study of fundamentals and of abstract concepts, is useful to the arts and concrete sciences. Precisely the theme, in other words, assigned to me for my lecture.

The *Protrepticus* was a well-known text in antiquity, cited by numerous authors. A consistent body of Aristotle's work has come down to us, but all from a much later period, written after he had left the Academy in Athens. He spends time on the island of Lesbos studying fish and other animals, and in the process founds the science of biology; he becomes tutor to the young future master of the world, Alexander, and to the group of friends who will later divide the empire between them and form the ruling families of the Hellenistic world, filling it with Aristotelian ideas and values. He returns to Athens, where he opens his own school, the Lyceum. The texts by Aristotle that have survived are probably the textbooks of the Lyceum, not one of them in dialogue form. The youthful dialogue on the usefulness of philosophy was lost in the cultural disaster that the conversion of the Roman Empire to Christianity entailed, with the systematic and brutal destruction of pagan thought, inaugurated by the Emperor Theodosius in the fourth century (with the subsequent destruction of the library of Alexandria, which can probably be attributed to Theophilus and his successor, St Cyril) and continuing unabated all the way to Justinian, who in 529 shut down the last incarnation of the Academy in Athens.

The modern reconstruction of the text of *Protrepticus* is an ongoing source of controversy. It is based mainly on an extensive work by Iamblichus, the Greek author of late antiquity who systematically copies and incorporates whole pages of work by the author whose ideas he is engaged in expounding. This makes it possible to put together from this work a plausible reconstruction of Aristotle's original dialogue. To prepare for my lecture, I read it. It turned out to be quite a surprise: the arguments used by Aristotle regarding the usefulness of philosophy for science are still completely relevant. All I had to do was copy them out myself and adjust them a little. Here they are.

The first argument is the most amusing, but it is also very subtle. Those who criticize the usefulness of philosophy for science, Aristotle has noticed, are not doing science: they are doing philosophy.

When Hawking and Weinberg do their great physics, they are scientists. When they write that philosophy is useless for science, they are not attempting to resolve a physics problem: they are merely reflecting on what may be considered useful, what methodology and conceptual structure is appropriate for doing science. Reflecting in this way is a useful enterprise and is precisely what philosophy does. The arrogantly pragmatic and 'anti-philosophical' attitude of Hawking and Weinberg, in fact, has its origins in . . . philosophy! We can easily trace it back to the philosophers of science who influenced their generation of scientists: the logical positivists with their anti-metaphysical rhetoric, followed by Karl Popper and Thomas Kuhn. Hawking and Weinberg are repeating ideas that come from the philosophy of science. Not only are they not aware of this, but they are not even up to date, since the philosophy of science has made some useful progress since Popper and Kuhn . . .

Aristotle's second argument is the most direct: the analysis of fundamentals has in fact had an influence on science.

If in the fourth century this might have seemed like wishful thinking, today it is an undeniable historical fact: the influence of philosophical thought on the finest Western science has been deep and persistent. Newton would not have existed without Descartes; Einstein learned directly from Leibniz, Berkeley and Mach, and from the philosophical writings of Poincaré; not to mention the fact that he would read Schopenhauer before falling asleep at night, and had read Kant's three *Critiques* before he was fifteen years old. The influence of positivism and of Mach upon Werner Heisenberg, the discoverer of quantum mechanics, is quite clear from his articles. And post-war American physics is inconceivable without the influence of pragmatism. And so on. A full list of this kind would be a very long one. Philosophical thought opens windows, frees us from prejudices, reveals incongruities and leaps of logic, suggests new methodological approaches, and in general opens up the minds of scientists to new possibilities. It has always done so in the past, and it continues to do so.

The reason that philosophical thought has this important role is the fact that the scientist is not a rational being with a fixed conceptual baggage who works on data and theories: he is a real being whose 'conceptual baggage' is continuously evolving as our knowledge gradually grows. Elaboration of general conceptual structure is what philosophers specialize in. It is above all in the area of scientific methodology, which is anything but fixed and static, that philosophy tends to interface with science. 'Philosophy,' writes Aristotle, 'offers a guide to how research should be conducted.'

The third of Aristotle's arguments is a simple observation: science needs philosophy 'especially where perplexities are greatest'.

When science undergoes periods of radical change during which fundamentals are questioned, it is most in need of philosophy. A prime example of this is our current moment, in which fundamental physics faces the problem of quantum gravity (on which I work), where our notions of space and time are once again under discussion, and the old debate on space and on time – from Aristotle to Kant, and all the way down to David Lewis – has become relevant again.

No, philosophy is hardly useless to science. It is, on the contrary, a vital source of inspiration, criticism and ideas.

But if the great science of the past was nourished by philosophy, it is also true that the great philosophy of the past was passionately nourished by science. Hume and Kant are incomprehensible without Newton. Or Descartes without Copernicus; or Aristotle without the Preocratic physicists; or Quine without Einstein's relativity. Even philosophers of the stature of Husserl and Hegel, who seem rather more distant from contemporary science, drew on the science of their time as a model of reference.

To shut one's eyes to contemporary scientific knowledge, as, alas, some philosophy in some European countries has done, is in my opinion simply ignorant. Even worse is the attitude of those currents in philosophy that consider scientific knowledge to be 'inauthentic', or of a lower order – or regard it as an arbitrary organization of thought that is no more effective than others. They remind me of two retired old men on a park bench: one mutters 'Scientists are so presumptuous. They think they can understand consciousness, or the origins of the universe!', and the other one grumbles: 'What arrogance. It's obvious they'll never succeed! To understand these things, of course, it takes . . . the two of us!'

Our knowledge is incomplete, but it is organic: it is constantly growing, and every part of it has influence over every other part. A science that closes its ears to philosophy fades into superficiality; a philosophy that pays no attention to the scientific knowledge of its time is obtuse and sterile. It betrays its own deepest roots, which are evident in the etymology of philosophy: the love of knowledge.

The Mind of an Octopus

During a boat trip some years ago, a friend of mine who was diving to hunt for octopus climbed back on board with nothing to show, and a slightly troubled air. 'There was an octopus in a hole,' he told us, 'but I didn't catch it because I lost my nerve: he was staring at me with his big eyes full of fear.'

A few days ago, the *Guardian* published a list of the ten most important books on the nature of consciousness. Number one, predictably enough, was Daniel Dennett's classic *Content and Consciousness*. But the second on the list was something of a surprise – *Other Minds* by Peter Godfrey-Smith, a book about octopuses. What on earth do these appealing marine creatures, all head and arms, have to do with consciousness?

Accounts like that of my friend are legion in the literature about octopuses. In the laboratories where they are being studied, scientists tell of octopuses capable of opening tins, of clandestinely escaping from their tanks and returning, closing the lid behind them; of recognizing the individual scientists in a research group and spraying the annoying ones with water; of working out how to short-circuit light bulbs when the light bothers them ... In natural environments, they have been observed behaving in complex and adaptable ways, and they seem to have the ability to recognize and interpret the attitudes and body language of those around them.

Octopuses have complex intellectual abilities that are decidedly unusual for creatures of their realm and are rather comparable in many respects with those of mammals. They have at their disposal an extremely rich and complex neural network. An octopus may have as many neurons as a dog, or a child. These are the characteristics that make them a valuable case study for those concerned with consciousness.

'Consciousness' is an ambiguous term that has come to mean various things. In the last few decades the phrase 'the problem of the nature of consciousness' has taken the place of what in the past used to be the problem of the meaning of soul, spirit, subjectivity, intelligence, perception, understanding, existing in the first person, being aware of a self . . . Not that these questions are equivalent – they clearly aren't – and what is meant by 'the problem of consciousness' changes from one author to another. But the question of how our subjective experience can arise from natural reality has taken central stage. One reason is that the existence of subjectivity remains the argument most insisted upon by those who, coming from various directions, resist a naturalistic perspective. What is it, in the great game of nature, this 'I' that I feel myself to be?

One way of tackling the issue is to observe our non-human cousins. If this doesn't provide us with all the answers, it can at least help to clarify the question. We have a lot in common with a cat or a dog, and a great deal more with chimpanzees. Hence there are a number of different questions. The first concerns the nature of the capacity to observe, to predict, to interact, to communicate, to suffer and to love – abilities and characteristics that we share with many mammals. The second, perhaps less interesting question concerns what it is, if anything, that differentiates our experience from that of our mammalian cousins. It is one thing to discover how the brain of a cat works; quite another to understand if and how the human brain works in a way that is different to that of a cat. As always, there is no better way of understanding ourselves than by comparing ourselves with others.

The brain and the behaviour of mammals, however, is too similar to ours, while if we stray too far, to remote biological relationships, we lose something essential to the comparison: we may be able to comprehend in depth how an amoeba works, but this does not give us the impression of having learned much about ourselves. Ideally, we would have an alien race to study, one that had arrived here from another planet, with elements of consciousness recognizably similar to our own but generated by different structures. Perhaps then we could grasp what is essential and what is an accessory to what we call consciousness.

For now at least, such aliens arrive only on the big screen, and these creatures tend to mimic humans in a rather unimaginative way. Sometimes the strangest thing about them is that they defend values that by sheer coincidence our civilization happens to be currently discussing. In truth, we are in a somewhat lonely place: we have nothing and no one with whom to compare consciousness or intelligence, beyond ourselves and our closest relatives. And this is where the octopus comes in.

The octopus is an extremely distant relative. The ancestors that we share with cats go back relatively few generations when compared to the gulf of many hundreds of millions of years that separates us from the ancestors that we have in common with the octopus. The process of separation has radically accentuated our differences, and the octopus belongs to a vast animal kingdom where signs of consciousness and intelligence like our own are quite rare. In that realm, they are exceptional: they have an extremely complex and rich nervous system with a similar number of neurons to mammals, even though they are distant from us in evolutionary terms, having evolved independently. Nature seems to have experimented with the creation of intelligence at least twice: once with our branch of the family, and a second time with the octopus. The octopus, in short, is the extra-terrestrial that we have been looking for in order to study a possible independent realization of consciousness.

Peter Godfrey-Smith is a philosopher concerned with the nature of consciousness, as well as a passionate scuba diver and a captivating writer. *Other Minds* is a work of popular science that describes the ingenious behaviours of these extraordinary creatures, and at the same time a convincing book about the nature of consciousness. It argues that consciousness is not something that does or does not exist: it is something that exists in different degrees and different forms: it is a form taken by the relations between an organism and the world.

What interests us about 'octopoid' intellectual complexity is not just the similarities with our own, it is also the differences between the two types. The neural structure of an octopus is different from ours: instead of being concentrated in a brain, it is articulated throughout its entire body, including its arms, diffused just below the surface of its body. It is a complex but radically alien intelligence. An octopus tentacle severed from its body continues to exhibit a complex capacity to process information.

An octopus has an amazing capacity to radically alter its skin colours and patterns, changing them rapidly. The colour of its epidermis is controlled by an extremely rich network of diffuse neurons, and the colour changes may also be used as a form of communication. I have no difficulty imagining what it must be like to be a cat. I watch a cat stretching out in the sunshine on a hot summer afternoon, and I can easily identify with that. But what must it feel like to be an octopus, with its brain spread throughout its body and its arms which can each think separately?

In the endless vastness of the galaxies, nature has in all probability given rise to every shape and form, making us one example among many. Who knows how many more complex forms are out there, partly similar to and partly different from ourselves, in the immense celestial expanses? Perhaps there is even one that swims in our seas. And the disturbing encounter that my friend had with the big, frightened eyes of the small octopus was nothing but the spark of an encounter between different kinds . . . of consciousness.

Ideas Don't Fall from the Sky

Years ago, during a physics conference, I found myself at dinner sitting next to the Nobel Prize winner Subrahmanyan Chandrasekhar, someone who for our generation of physicists had a mythical status on account of his creativity. 'Chandra' was at the time an elderly and affable gentleman of few words. In the middle of the meal he turned to me and said: 'You know, Carlo, in order to do good physics . . .' My eyes widened, and I froze in anticipation of some priceless, oracular gem. '. . . in order to do good physics, what is needed most is not to be very intelligent.' Coming from the brilliant scientist who had understood the upper limit of the mass of stars, and who had developed the mathematical theory of black holes, this was an idea that sounded absurd. But what followed, in conclusion, was disarming: 'What matters most is to work very hard.'

I am reminded of Chandrasekhar's words every time I come across some instance of the myth of 'pure creativity' or of 'unfettered imagination'. To construct the new, I have heard it said, it is enough to violate rules and liberate oneself from the dead weight of the past. I don't think creativity in science works like this. Einstein did not just wake up one morning thinking that nothing was faster than light. Nor did Copernicus simply think up the idea that the Earth orbits the sun. Or Darwin that species evolve. New ideas do not just fall from the sky.

They are born from a deep immersion in contemporary

knowledge. From making that knowledge intensely your own, to the point where you are living immersed in it. From endlessly turning over the open questions, trying all roads to a solution, then again trying all the roads to a solution – and then trying all those roads again. Until there, where we least expected it, we discover a gap, a fissure, a way through. Something that nobody had noticed before, but that is not in contradiction with what we know; something minuscule on which to exert leverage, to scratch the smooth and unreliable edge of our unfathomable ignorance, to open a breach on to new territory.

This is the way that most creative minds in science have worked, and how thousands of researchers work today, in order to advance our knowledge. Ideas are disclosed in a long and unnerving traffic with the margins of our knowledge.

Copernicus was familiar with Ptolemy's old book, down to the last detail, and in its folds he glimpsed the new shape of the world. Kepler struggled for years with the data gathered before him by the astronomer Tycho Brahe, before deciphering amongst those data the elliptical orbits that provided the key to understanding the solar system.

New knowledge emerges from present-day knowledge because within it there are contradictions, unresolved tensions, details that don't add up, fracture lines. Electromagnetism was difficult to fully reconcile with Newton's mechanics – and this provided Einstein with an opportunity. The elegant elliptical trajectories of the planets as discovered by Kepler could not be made to square with the parabolas calculated by Galileo, and this provided Newton with the key to moving forward. Atomic spectra that had been measured for years could not be made to fit classical mechanics, and this provoked Heisenberg no end. The internal tensions between one theory and another, between data and theory, between different components of our knowledge, generate the apparently irresolvable tensions from which the new springs. The new breaks the old rules, but in order to resolve contradictions rather than for the sake of it.

In a tremendous passage of his *Letter VIII*, Plato gives an account of the process of acquiring knowledge:

> After many efforts, when names, definitions, observations and other sensory data are brought into contact and compared in depth, one juxtaposed with another, in the course of a scrutiny and an even-tempered but severe examination, at the end a light suddenly comes on, for whatever problem – our understanding, and a clarity of intelligence the effects of which express the limits of human power.

Clarity of intelligence . . . but only after 'many efforts'.

Two thousand four hundred years later, Alain Connes, one of the greatest living mathematicians, describes the discovery of what makes one a mathematician in the following words:

> One studies, continues to study, studies still, then one day, through study, a strange sensation surfaces: but it can't be, it can't be so, there is something that does not work out. At that moment, you are a scientist.

The Many Errors
of Einstein

There can be no doubt that Albert Einstein was one of the greatest scientists of the twentieth century, one who saw deeper into the secrets of nature than anyone. Does this mean that we should take everything he did as correct? That he never made mistakes? On the contrary.

In fact, few scientists have made as many errors as Einstein. Few have changed their minds as frequently as he did. I'm not talking about the kind of mistakes he made in his everyday life, which are a matter of opinion and, ultimately, his own business. I'm talking about genuine scientific errors: mistaken ideas, wrong predictions, error-strewn equations, scientific assertions that he himself came to regret and that were proved false.

Let me give you a few examples. Today we know that the universe is expanding. The Belgian physicist Georges Lemaître managed to understand this by using Einstein's own theories, and informed him of his findings. Einstein responded by dismissing the ideas as nonsensical, only to have to eat his words when in the thirties the expansion was actually observed. Another consequence of his theory is the existence of black holes, and he wrote several erroneous texts on the subject, contending that the universe ends at the edge of a black hole. The existence of gravitational waves, for which we now have good indirect evidence,* also followed from Einstein's theories. Einstein wrote at first that these waves existed, but only before claiming that they did not – in effect misinterpreting his own theory – and then changing his mind again to accept the opposite, correct conclusion.

When he wrote his theory of special relativity, Einstein did not use the notion of *spacetime*. This notion, which is to say the concept of a four-dimensional continuum that includes both space and time, is really down to Hermann Minkowski, who used it to rewrite Einstein's theory. When Einstein became aware of what Minkowski had done, he maintained that it was merely a useless mathematical complication of his theory – only to change his opinion completely, shortly afterwards, and use precisely the concept of *spacetime* in order to write the theory of general relativity.

On the issue of the role of mathematics in physics, Einstein repeatedly shifted his point of view, advocating in the course of his life various ideas that were in direct contradiction with each other.

Before writing the correct equations of his major work, the general theory of relativity, Einstein had published a series of articles, all wrong, each proposing a different incorrect equation. He even went so far as to publish a complex and detailed work to argue that the theory cannot have a certain symmetry . . . that he later chose as the foundation of his theory!

In the final years of his life, Einstein obstinately persisted in wanting to write a unified theory of gravity and

* The direct observation of gravitational waves was achieved on 14 September 2015, five months after the publication of this article. It was followed in 2017 by the Nobel Prize for physics.

electromagnetism, without realizing, as it would shortly be understood, that electromagnetism is a component of a larger theory (electroweak theory), and therefore that his project of unifying it with gravity was pointless.

Einstein also shifted his position repeatedly in the great debates about quantum mechanics. At first, he argued that the theory is contradictory. Then he accepted the idea that it wasn't, confining himself to insisting that it must be incomplete and that it failed to describe all of nature.

Regarding general relativity, for a long time Einstein was convinced that the equations could not have solutions in the absence of matter, and therefore the gravitational field depended on matter – only to then change his mind when Willem de Sitter and others showed that he was wrong, ending up by interpreting the gravitational field as a separate, real entity that exists in its own right.

In the extraordinary work of 1917 in which he founded modern cosmology, Einstein understood that the universe can be a three-sphere and introduced the cosmological constant which has today been verified, and managed to add together an egregious error of physics – the idea that the universe must not change in time – and a resounding error in mathematics: he didn't realize that the solution he wrote was unstable, and that it could not describe the real universe. As a result, the article is a strange mixture of major new and revolutionary ideas and a mass of serious errors.

Do all these mistakes and changes of opinion take something away from our admiration of Albert Einstein? Not at all. If anything, the opposite is the case. They teach us something instead, I believe, about the nature of intelligence. Intelligence is not about stubborn adherence to your own opinions. It requires readiness to change and even discard those opinions.

In order to understand the world, you need to have the courage to experiment with ideas not to fear failure, to constantly revise your opinions, to make them work better.

The Einstein who makes more errors than anyone else is precisely the same Einstein who succeeds in understanding more about nature than anyone else, and these are complementary and necessary aspects of the same profound intelligence: the audacity of thought, the courage to take risks, the lack of faith in received ideas – including, crucially, one's own.

To have the courage to make mistakes, to change one's ideas, not once but repeatedly, in order to discover. In order to arrive at understanding.

What's important is not being right. It's to try to understand.

Some Think, O King Hiero, That the Grains of Sand Cannot be Counted

Among the many cultural roots of the world in which we live, rational Greek thought and the texts of the Bible are two of the most influential. Since late antiquity, strenuous attempts to reconcile these two strands have been made by Christian intellectuals, with mixed results, and still today their relationship with rational thinking is an issue for Christianity and Islam. Is it possible that elements of this dialogue were already in place two centuries before the birth of Christ? The idea is a small gem that can be found in Giuseppe Boscarino's recent critical edition and translation of *Psammites* (*The Sand Reckoner*) by Archimedes.

Archimedes, who lived in Sicily immediately before its occupation by the Romans, is one of the great scientific figures of antiquity. From Archimedes we have inherited pages of stunning mathematical richness that played an important part in the renaissance of scientific thought in the modern era and continued to inspire the development of mathematics right up until the end of the nineteenth century.

From the writings of Archimedes this strange little book, *The Sand Reckoner*, in which he literally counts grains of sand, has also survived. Counting grains of sand seems like an undignified occupation for a scientist. What is he up to?

The Sand Reckoner is not really a work of science. It is really more of a 'popular' text, as we deduce from a reference to more technical, now lost, work by the same Archimedes.

The problem *The Sand Reckoner* deals with is the construction of an arithmetical system. The numbers used in the third century BC do not allow us to count very numerous things. The numerical system was the Greek one, similar to the Roman in which X was 10, V was 5, and 15 was written XV. The largest number to have a specific name was 10,000, which was called a 'myriad' and indicated in Greek with the letter M. There was no way of writing directly, and hence no way of using, numbers that were much bigger than this. Archimedes tackled the problem and constructed a system that could deal with arbitrarily large numbers. The solution is to call a myriad of myriads 'a second order number'. In this way two second-order numbers are two myriads of myriads, which is to say 200 million. A myriad of myriads of numbers of the second order results in a number of the third order, which is to say 10 million billion, and so on. It is a solution similar to the one in modern science: we use the powers of ten.

In *Psammites*, Archimedes gives a demonstration of the usefulness of this system by estimating the total number of grains of sand in the world. In fact, he goes one better than this: he estimates how many grains there would be in the entire universe, if the universe happened to be filled with sand.

First he estimates how many grains of sand would fill a mustard seed, then how many mustard seeds would fill a box the size of his index finger, how many boxes would fill the Earth, how many Earths the solar system, and finally how many solar systems would fit into the universe, according to the astronomy of the time.

During the course of this calculation he reveals the technical accuracy at his disposal to measure the diameter and the distance from us of the sun and the moon, and much else besides about the astronomy that he is familiar with.

Particularly fascinating is his reference to the heliocentric theory of Aristarchus, which anticipates Copernicus by some fifteen centuries. The final outcome of the reckoning is that the number of grains of sand needed to fill the universe is a thousand myriads of numbers of the eighth order, which is to say, in modern terms, ten to the power of sixty-three (10^{63}). A big number, but one that is definite and conceivable.

The game is a refined one, and the playful execution of it by Archimedes is carried out impeccably. Beneath the playfulness you also get the sense that there is something essential at stake. At the beginning of the text, couched in the form of a letter, its polemical objective is made quite explicit: 'There are those who think, O King Hiero, that the grains of sand cannot be counted.' And this reminds us of a passage in the Bible:

Who can number the sand of the sea, and the drops of rain, and the days of eternity? Who can find out the height of heaven, the breadth of the Earth, the depth of the abysses? [. . .] There is only one who has this wisdom: the Lord sitting upon his throne.

These are the powerful opening words of the book of *Ecclesiasticus*, or *Sirach*. They speak of counting grains of sand, but emphasize the impossibility of such reckoning and such knowledge. Can there be a connection between the two texts?

Ecclesiasticus was probably written in Palestine in the Hellenistic era, which is to say in a period of Greek political and cultural domination, and then translated soon afterwards into Greek in Egypt, as is mentioned in the text itself, probably in Alexandria, where the Greek dynasty of the Ptolemies was involved in an effort to compile, translate, study and conserve the entire knowledge of antiquity. It is largely thanks to this effort that the Bible is available to be appropriated by the Jewish, Christian and Muslim traditions. The Bible that we are familiar with, in other words, was compiled and edited in Alexandria under the initiative of enlightened Greek sovereigns: it was thanks to Greek universalism and multiculturalism that the Bible was transmitted to us, more than to the particular culture of the ancient Jewish world. In that same Alexandria, in the public institutes of research, the celebrated Library and the Museum (prototypes of the modern university), a brilliant young Sicilian – Archimedes – probably studied, and kept in epistolary touch with the Alexandrine intelligentsia for the rest of his life.

On close inspection, the translator of *Ecclesiaticus* into Greek mentions that he came across the book in the thirty-eighth year of the reign of Euergetes, which is to say 140 years before the birth of Christ – at a time, that is, when Archimedes had already been killed by the Romans. But perhaps we should take these dates with a pinch of salt, since Archimedes could have come into direct contact with the Hebrew text in Alexandria, or with similar Hebrew texts. In Alexandria it had been customary for over a century to systematically translate learned Hebrew works into Greek. The conceit of the grains of sand that cannot be counted, rhetorically standing for irrevocable human limitations, existed prior to *Ecclesiasticus*. Pindar tells us, for instance, centuries earlier, that 'Sand escapes from counting.'

Taking all this into account, perhaps the polemical objective of Archimedes starts to become clear. With an enlightenment flourish before the letter, Archimedes rebels against the type of knowledge that insists on mysteries that are intrinsically beyond human understanding. Archimedes does not pretend to know the exact dimensions of the universe, or the precise number of grains of sand. It isn't the comprehensive nature of his own knowledge that he is defending. On the contrary, he is quite explicit about the approximate and provisional nature of the estimations that he makes. He talks, for instance, of various alternatives regarding the dimensions of the universe, a subject on which he does not have a well-defined position. And he is aware in fact of how yesterday's ignorance may be enlightened today, and how today's knowledge may be revised tomorrow.

But he rebels against abandoning the search for knowledge. His is a declaration of faith in the knowability of the world, and a proud rebuke to those who are content with their own ignorance and with delegating knowledge elsewhere.

Many centuries have passed, and the text of *Ecclesiasticus*, along with the rest of the Bible, can be found in countless homes all over the planet, while Archimedes' text is read only by a few. Archimedes was slaughtered by the Romans during the sacking of Syracuse, the last proud remnant of Magna Grecia to fall under the Roman yoke, during the expansion of that future empire that would soon adopt *Ecclesiasticus* as one of the foundational texts of its official religion, a position which it was to occupy there for more than a thousand years. During that millennium, the calculations made by Archimedes languished in a state of incomprehensibility. Near to Syracuse there is one of the most beautiful sites in all Italy, the theatre at Taormina, overlooking from above the Mediterranean Sea and Mount Etna. In the time of Archimedes, the theatre was used to stage plays by Sophocles and Euripides. The Romans adapted it for gladiatorial combat. In other words, the cultural battle between the world of *Ecclesiasticus* and that of Archimedes saw the complete triumph of the former.

But let's look again, point by point, at the text from *Ecclesiasticus*. The number of grains on the shores of the seas was estimated by Archimedes; the number of drops of rain enters into the calculations of climatologists; the number of days since the Big Bang have been determined by cosmology; Aristarchus had already begun to measure the height of the sky; the extension of the Earth had been assessed by Eratosthenes some decades before Archimedes, and today it is known, like the depth of marine abysses, with millimetrical precision. For the supposedly unanswerable questions listed in *Ecclesiasticus*, we have found answers.

In the meantime, new open questions have arisen.

The issue posed by Archimedes is still current: do we want to go in search of what we still don't know, or should we simply accept the idea that our knowledge has fixed boundaries?

The subtle, intelligent game played in *The Sand Reckoner* is not just a demonstration of an audacious mathematical construction, or the virtuosity of one of the most extraordinary intellects of antiquity. It is also a defiant cry of reason, which recognizes its own ignorance but refuses to delegate knowledge to others. It is a small, low key and extremely intelligent manifesto against obscurantism. It has never been more current.

Why Does Inequality Exist?

When Europeans first came into contact with populations of 'primitive' people, they thought that the lifestyle of these people must have remained the same as it had been through long millennia of prehistory. This idea was later criticized for its naïvety, but today it is finding new supporters. This comes out of a vast interdisciplinary collaboration between anthropology and archaeology.

Decades of work by anthropologists who have lived with indigenous populations on every continent has furnished ample and detailed descriptions of their ways of life and culture. At the same time, decades of archaeological digs have provided us with an increasingly precise picture of prehistoric life. Comparison between the two is illuminating. We find in both similar objects, similar amulets, similar structures of houses and villages, similar statues and monuments such as dolmens, similar forms of sustenance, commercial exchange, similar rituals. And so on. It is then reasonable to think that the 'primitive' populations studied by anthropologists can indeed provide a window on to the way of life of our species during different phases of the Stone Age. Looking through their eyes is perhaps like looking through the eyes of thousands of generations that came before us.

This is the starting point of *The Creation of Inequality: How Our Prehistoric Ancestors Set the Stage for Monarchy, Slavery and Empire*. The book is co-authored by an archaeologist and an anthropologist: Kent Flannery and Joyce Marcus, who are known for their important contributions to the study of the Pre-Columbian cultures of Central America.

It culminates in a surprising thesis on the origins of inequality in human society – one that has considerable political and social resonance. Many societies, including our own democracy, are rigidly stratified: billionaires and the poor; aristocrats and commoners, generals and privates, freemen and slaves, etc. What is the origin of this widespread inequality? Was the human species always so hierarchical? Classic political thought is divided on the subject, with a spectrum of contrasting theories. From the divine origin of inequality, which purports to show that nobles and bourgeois Calvinists, the king and the Pope are above others by virtue of divine grace – to the celebrated and frequently ridiculed thesis in *A Discourse on the Origin and Foundations of Inequality among Men*, published by a young Jean-Jacques Rousseau in 1755, which proposes the idea that primitive society was an egalitarian one in which all men and women had equal dignity and resources were shared equally. According to Rousseau, this ideal state of 'noble savagery' was lost with the structuring of society that brought about the formation of social classes, powers and inequalities. Rousseau's powerful words to the effect that 'Man is born free, and everywhere he is in chains' were published shortly before the Revolution in France.

Today, recent research provides evidence in support of Rousseau. Before the invention of agriculture, before the formation of complex social structures such as tribes and clans, our ancestors lived as hunter-gatherers, organized into small groups in which social equality was actively defended.

The basic structure of ancient, nomadic, hunter-gatherer society was the extended family, formed of ten to twenty individuals with close blood ties, and linked in turn to other extended families living in the same area, through a dense web of gift exchanges. There is no accumulation of wealth, and no 'class' differences. Even recognition of the outstanding skills possessed by an individual, in hunting for instance, was reined in by the culture. Among the Kung of the Kalahari Desert between Namibia and Botswana, for example, a particularly successful hunt by an exceptional hunter will be met with a joyful commotion, but also with a certain amount of debunking irony: a rich haul of meat immediately attracts mocking commentary describing it as 'a useless bag of skin and bones', etc. The society is keen to ensure that no one should feel or find themselves in a position of privilege of any kind.

The spoils of the hunt are immediately divided up and distributed. A man who is more than averagely capable and intelligent may be listened to from time to time if he manages to convince others to go with a decision of his, but the group has no leader, and anyone seeking to dominate would immediately be ostracized by the group. The only wealth that a family accumulates is credit with its neighbours, through making many gifts. In times of difficulty, those neighbours will be happy to repay the favour. This seems to have been the basic way of life of humans for hundreds of thousands of years.

But around fifteen thousand years ago, with the growth in ability to cultivate the fruits of the earth, the consequent rise in population and the need to work together in large groups, a new structure began to develop in which social distinctions developed and the value of equality weakened. The structure that became most prevalent was that of the clan, formed of numerous extended families. The clan creates a new identity that is neither strictly familial nor dependent on gift exchange. It may recognize itself in some mythic ancestor, but it is founded on a specific institution where young men struggle arduously to belong and become gradually initiated into the secrets of the group. The 'men's house', the central building in the village, is a place of socialization where young men are initiated and educated in the values of the group. It is what we find, in almost identical forms, in indigenous villages on every continent, and is matched by the archaeological excavations that take us back to the period of transition from the primitive phase of hunter-gathering. The 'men's house' is the origin of many modern institutions, from churches to schools, from barracks to universities. The life of the clan revolves around this place and is rooted in complex rituals that transmit its founding myths.

It is in this phase that the conspicuous success of certain individuals begins to be socially recognized. It is here also that men begin to assign higher value to their own gender, considering themselves to be the backbone of the society, and to therefore marginalize women; and older men who are already initiated into the society acquire status in relation to the young, who must go through and complete the process. The clan is run by a successful minority that controls its rites and initiation ceremonies and has custody of its secret knowledge. It is the origin of aristocracy, of the clergy and of large concentrations of wealth. Inequality within human society is born.

In a second phase that begins seven and a half thousand years ago in the Near East, four thousand years ago in Peru and three thousand years ago in Mexico, a privileged elite actively organize themselves to stabilize their powers and make them hereditary. This is the phase in which conflicts between villages evolve into wars of conquest, religion reorganizes itself and, according to our authors, the men's house becomes a temple, a place of worship specifically dedicated to the cult of a major divinity who ratifies these privileges. Various societies oscillate between phases in which the elite prevails and centralizes power, and others in which the majority re-establishes the values of equality. The Konyak Naga in the mountains of Assam, between India and Bhutan, for instance, were observed shifting cyclically between a more egalitarian structure based on the recognition of individual merit, called *thenkoh*, and one based on rank, with hereditary leaders, called *thendu*.

This brief sketch hardly does justice to the wealth of vivid descriptions of the lives of people from every imaginable corner of the globe that are presented in this book. Festivities that could not do without, for good luck, a raid against the neighbouring village to cut off a few heads (an apparently quite widespread custom . . .). Battles of escalating gift-giving to acquire prestige (I give you a pig, you give me two, I give you three . . .), until the stage is reached where the person no longer able to reciprocate submits to a form of servitude. The book is a treasure trove of fascinating stories. Anyone interested in the ancient route that has led to what we have become today should read it.

We should not, of course, read too much into the results of research that is still ongoing. There is still so much to be found out about our past. But it is difficult to resist the compelling idea that human history contained an extremely long period in which a kind of egalitarianism was the norm.

The impetus towards a society where inequality is kept in check has been profoundly rooted in our civilization since the epic royal hunts of the archaic Greco-Roman period. In the first Roman republic the Licinian Rogations, laws instituted in 367 BC, limited the amount of wealth (land and livestock) that the richest patricians were allowed to possess. The desire for equality has marked the growth of the modern world: from the abolition of slavery to the abolition in the eighteenth century of the privileges of the aristocracy and clergy, right up to the modern idea of a democracy in which each vote has equal value. The recent historic failure of real socialism has arrested this impetus, and today, barely veiled by pro-democracy rhetoric, we are witnessing in every corner of the globe an almost ferocious radicalization of inequality: the distribution of wealth is becoming ever more unbalanced in every country, and the world has witnessed the emergence of a super-rich elite in which power is concentrated.

The nineteenth- and twentieth-century ideal of equality, still vividly alive just a few decades ago, is today faded and derided. Perhaps it is just part of an oscillation, like those of the Konyak Naga in the mountains of Assam. Deep within our cultural genes there are perhaps tens of thousands of years of a society which was hardly ideal but which sought to distribute its resources equally. Where all its men and women were considered equal.

Dramatic Echoes of Ancient Wars

A spectacular archaeological find, in an area called Nataruk on the ancient banks of Lake Turkana in Kenya, shows the still-harrowing evidence of an episode of war that took place there ten thousand years ago. News of it has been published in the magazine *Nature*, by a group of English, Kenyan, Australian and Indian archaeologists. The importance of the find lies in the light it may shed on a desperately urgent question: why do we wage war?

There are two main theories concerning the origin of warfare. The first maintains that war is relatively recent in origin: it arose with the birth of agriculture, when people had begun to accumulate resources – harvests stored in granaries, for instance. These resources became attractive to other groups, triggering cycles of robbery and violence. Two factors support this hypothesis. Firstly, populations that today follow a way of life that is pre-agricultural, pre-cattle-raising and pre-sheep-farming do not, in general, wage war. They live by hunting and gathering, as humans have done for thousands of years. This is a form of subsistence that does not permit accumulation: every surplus is perishable, and it is best to give it away, receiving gratitude and recognition in exchange. In populations such as this, encounters with other groups are occasions for gift-giving and receiving, and for the formation of new couples. The second observation in support of the

81

recent origin of warfare is the absence of violence in one of the two species that most resemble us: the bonobo. When groups of these small, lively, West African chimpanzees meet, it becomes a festive occasion.

The opposite theory is that violence between groups is intrinsic to our species, and that war has existed throughout the long prehistory of *Homo sapiens*, which is to say for millions of years. In support of this second view, there is the behaviour of another species close to our own: the common chimpanzee. Violent confrontations between groups of these apes is commonplace and can lead to the killing of enemies.

Which version is right? Throughout the hundreds of millennia of our prehistory, when we wandered nomadically about the world in small groups, hunting with bows and arrows and collecting herbs, berries and roots, were we pleased or terrified when we came across a group of the same species as ourselves? Were we thinking of how to thrust a pre-emptive spear into their bellies, or was it an occasion for making gifts, and for young men and women to exchange meaningful glances with each other?

The discoveries at Lake Turkana add something to the debate. Ten thousand years ago, at Nataruk, on the ancient shore of a lake that was much bigger at the time, there was a massacre. All the evidence suggests that it was a massacre of one group of humans by another. It is the oldest evidence of conflict between humans that we have found to date. Twenty-seven skeletons of men, women and children have been found, in a variety of postures, without ritual burial and with obvious signs of violence having been done to them: crushed skulls, stone arrowheads in ribcages, indications that some of them may have been tied up, evidence of massive traumas and fractures to the arms and legs of victims. There are signs that maces, arrows or spears were used. One man has an obsidian blade still stuck in his skull, another a wound in his cranium from a blow that crushed part of his face, and an arrowhead in his knee. He fell headfirst into what had probably been a low-lying lagoon. A pregnant woman is in a contorted position that suggests that her hands and feet had been bound and then tied together. The site has yet to be fully excavated, so the victims of this slaughter could turn out to be more numerous. From the remains of what in all likelihood were the weapons used to perpetrate it, there are blades made of obsidian from another geographical area, indicating that at least one of the groups that came into conflict on the banks of Lake Turkana may have come from elsewhere. What we are dealing with are dramatic echoes of ancient wars.

The discovery puts in place an element of evidence about the origins of war that seems certain: ten thousand years ago in East Africa, before the great Neolithic revolution that spread agriculture and made possible the birth of civilizations, there already existed violent conflicts between groups of humans, and these conflicts could culminate in massacres.

Does this mean that war has always existed? Is the hope we have of putting an end to the barbarism of war just the noble product of civilization – and a very recent one at that?

Perhaps not. On the contrary, the archaeologists who studied the site emphasize that the discovery might be part of a case proving that war itself is recent. Ten thousand years ago, the western shores of Lake Turkana formed a particularly fertile coast that would have been able to sustain a high density of groups of hunter-gatherers. There is evidence there of the use of ceramic vessels that may indicate that reduced nomadism and the accumulation of primitive resources had already begun. Ten thousand years ago is not so far distant from the erection of the first Pyramids. If these clues are confirmed, and it becomes clear that there is no evidence of violent conflict *preceding* this period in the history of humanity, then the discovery will serve to prove that war is a recent development. To prove that we humans did not evolve over millions of years to be as bestial as those ancient skeletons and our daily news seems to show. The disgust for war that many of us feel may be rooted in the instinctual mental fabric of our species.

A Scandal That Has Lasted for Ten Thousand Years is the subtitle of what I believe is one of the greatest Italian novels: *History*, by Elsa Morante. And 'ten thousand years' is precisely the age of the find at Nataruk, almost as if Morante already knew about it, with her intense humanity and her clear-eyed and moving vision of life. Ten thousand years is a long time: too long for the despair of countless victims such as her Ida Ramundo, the mother of the unforgettable Useppe, or for the pregnant woman bound and killed at Lake Turkana, or for the Ida Ramundos of today, cowering beneath the bombs falling on Aleppo. But perhaps we should also inflect Elsa Morante's terms to think of the duration of this scandal as *only* ten thousand years, rather than millions. A long scandal, but one that it is perhaps not too late to extricate ourselves from.

Four Questions for Politics

Soon an electoral campaign will be under way in Italy. I will vote for the party that seems to me most credible when addressing four problems I believe to be the most serious.

The first is the spread of war, causing extreme suffering, refugees and instability. The second is climate change and the other ecological and medical emergencies that are putting the future of our species at risk. The third is the current breakneck increase in economic inequality, and the kinds of concentrations of wealth that are immoral and that generate conflict. The fourth is the presence of vast atomic arsenals, which continue to represent a real and terrible risk, heightened by various recent threats to use them.

These seem to me to be serious risks to us all. Only politics can resolve them; but it can only do so if we as citizens reward a political force that wants to tackle them.

In foreign policy, Italy is one of the world's major industrialized countries, and with the exit of the United Kingdom it is one of the three principal countries of the European Union. It does not need to always follow in the slipstream of other Western countries. It can make its presence felt, use its voice to communicate values and offer concrete proposals in relation to each of these problems. It can assume a clear-cut position against war, withdrawing the many troops that it has positioned around the world, placing them at the disposal of the United Nations instead.

Internally, Italy can stop contributing to the slaughter, as it is currently doing as one of the major exporters of armaments. Innocent families have been annihilated in Yemen, on a daily basis, with bombs made in Sardinia.

It can take advanced measures to reduce carbon emissions and the promotion of environmentally friendly political aims, as other far-seeing states have done.

It can seek to even out the social disparities within its system, in the most straightforward and traditional sense, by taxing those with the greatest wealth. The redistribution of wealth is a principal function of the state, and the current growing concentration of wealth is both extreme and dangerous.

With a justified sense of pride, it can simply liberate itself from all the nuclear weapons belonging to others that are currently on its territory. It has this right – and the moral duty towards itself and to others.

These are four major objectives, and they are not unrealistic: in the past, humanity knew how to find ways of repeatedly curbing the outbreak of war, to agree on measures to safeguard the environment, to correct exaggerated social inequalities and scale down the nuclear arsenal. There is no reason why we cannot do so now.

But to do so requires an engagement from each one of us, in the simplest form possible: voting for a party that wants to do so.

Each one of these objectives has an inevitable political cost, because each will upset some people. I will vote for the party that will undertake to implement these objectives.

I believe that the future can be better for all of us and

the worst dangers avoided only if common interests prevail over those of individuals, and if collaboration prevails over conflict. If talk prevails over force, and dialogue with others over fear of others. Within our own country and throughout the world. I am waiting for a party speaking this language.

National Identity is Toxic

Great Britain is an old country. My own country, Italy, is relatively young. Both are proud of their past. Both have marked national characteristics: it is easy to identify the Italians and the British among the crowds at international airports. I recognize the Italian in myself: I can't say anything without waving my hands around, there are ancient Roman stones in the cellar of my house in Verona, and my schoolboy heroes were Leonardo and Michelangelo . . .

And yet this national identity is only a thin layer, one among others that are far more important. Dante shaped my education, but so too, and to a greater extent, did Shakespeare and Dostoevsky. I was born in conservative, bigoted Verona, and going to study in libertine Bologna was something of a culture shock. I was brought up within a particular social class, and I share the habits and preoccupations of people of this class everywhere else on the planet, more than with fellow Italians in general. I belong to a particular generation: an Englishman of my age has more in common with me than a Veronese of a different generation. My identity comes from my family, which is unique – just as every family is unique – from the groups of friends I had growing up, from the cultural tribe I chose in my youth, from the network of geographically scattered friends made in my adult life.

It comes, above all, from a constellation of values, ideas, books, political dreams, cultural preoccupations, common objectives that were shared and nourished, for which we struggled together, and which were transmitted in communities that were bigger, or smaller, and that completely cut across the confines of the nation. This is what we all are: a combination of layers, intersections in a web of exchanges woven by the whole of humanity in its multiform and ever-changing culture.

I'm only stating the obvious. So why, if this is the variegated identity of each one of us, do we organize our collective political conduct into nations and base our sense of identity on belonging to a nation? Why Italy? Why the United Kingdom?

The answer, once again, is simple: it isn't power that constructs itself around a national identity, it's the other way around: national identity is created by the structure of power. Seen from the perspective of my young and still somewhat dysfunctional country, Italy, this is perhaps easier to notice than from within the very old and noble realm of Her Majesty the Queen. But the same applies. As soon as it emerges, usually through fire and fury, the main preoccupation of any centre of power – be it a novel monarchy or a liberal bourgeoisie of the nineteenth century – is to promote a robust sense of shared identity. 'We have made Italy, now we must make Italians!' is the famous exclamation attributed to the pioneer of Italian unification, Massimo d'Azeglio.

I am always surprised by how differently history is taught in different countries. For someone from France, the history of the world revolves around the French Revolution. For someone raised in Italy, the (Italian) Renaissance and the Roman Empire are the events of universal significance. For an American, the crucial event for humanity, the one that ushered in the modern world, democracy, and liberty itself, was the war of Independence against . . . Great Britain. For an Indian, the roots of civilization are to be found in the time of the Vedas . . . Everyone finds slightly ridiculous the distortions of history caused by the national perspective of others. No one reflects upon their own.

We tend to read the world in terms of grand narratives of discord and struggle, which we share with our countrymen. These are narratives deliberately created to generate a sense of belonging to fictitious families, called nations. Less than two centuries ago there were people in Calabria who called themselves 'Greeks', and not so long ago the inhabitants of Constantinople defined themselves as 'Romans' . . . and not everyone in Scotland and Wales supported England in the last World Cup . . . National identities are nothing more than political theatre.

Don't get me wrong. I'm not saying that there is something necessarily bad about all of the above. On the contrary: to unify a diverse population – Venetians and Sicilians, say, or various Anglophone tribes – so that they can collaborate for the common good, is wise and far-sighted politics. If we fight amongst ourselves, it is obviously much worse than if we work together. It is cooperation, not conflict, that benefits everyone. The whole of human civilization is the result of cooperation. Whatever the difference between Naples and Verona, things are better for everyone without a border between them. The exchange of ideas and merchandise, gazes and smiles, the threads that weave the fabric of civilization, enriches everyone in goods, intelligence and spirit. Helping diverse people to converge in a common political space is to everyone's advantage. To then reinforce this process with a little ideology and political theatre, to keep internal conflicts at bay, to put on the farce of Sacred National Identity, is a useful strategy, regardless of how fake it is. It takes people for a ride, but who can deny that cooperation is better than conflict?

But it is at precisely this point that national identity becomes poisonous. Created to foster solidarity, it ends up becoming an obstacle to larger-scale cooperation. Forged in order to reduce internal conflicts, it ends up provoking external ones that are even more damaging. The founding fathers of my own country had good intentions when promoting the idea of an Italian national identity, but only a few decades later this resulted in fascism, the extreme glorification of national identity. Italian fascism inspired Hitler's Nazism. The passionate emotional identification of the Germans with a single *Volk* culminated in the devastation of Germany and a good part of the world. When the national interest promotes conflict rather than cooperation, when instead of searching for compromise and agreed rules it is thought preferable to put one's nation above all else, national identity becomes toxic.

Nationalist and sovereignty-obsessed politics are spreading throughout the world, increasing tensions, planting the seeds of conflict, threatening all of us. My own country has just fallen prey to this recklessness. I think that it's time to say loud and clear, in response, that national identity is a con. It's good for overcoming local interests for the common good, but it is short-sighted and counterproductive when it promotes the interests of a totally artificial group – 'our nation' – above a more ample sense of what the common good consists of.

But localism and nationalism are not only miscalculations, they draw power from their emotional appeal: they offer an identity. Politics likes to play upon our insatiable desire to belong. 'Foxes have dens, and the birds of the air have nests, but the Son of Man has no place to lay his head . . .' The nation offers just such a home, a fictitious one: it is a fake, but is costs little and pays politically.

Not that we don't have national identities: we do have them. But each of us is a crossroads of multiple, layered identities. Putting the nation first means betraying all the others. Not because we are all equal in the world, but because we are different within each nation. Not because we don't need a home, but because we have better and more dignified homes outside the grotesque theatre of nationalism: our families, our fellow-travellers, the community we share values with, spread throughout the world. Whoever we are, we are not alone: we are many. And we have a wonderful place to call 'home': the Earth, and a marvellous, variegated tribe of brothers and sisters with whom we may feel at home and identify with: humankind.

Charles Darwin

Throughout history we have made some fundamental advances in our understanding of the world. Big steps forward after which there was no turning back. Among the most important of these is the one made in the mid-nineteenth century by Charles Darwin.

Darwin's discovery relates to all the living beings on this Earth: from mice to butterflies; from viruses to elephants. As well as, dear reader, to you and me. The first thing that we understood thanks to Darwin's work is that all living beings share the same ancestors. We are part of the same genealogy of the one great family to which we all belong. The mother of the mother of the mother of a butterfly in your garden is also the mother of the mother of the mother . . . of your mother. If you think about it, this discovery is both thrilling and moving. We are all siblings on this planet. And this is a fact, of the same order of certainty as that the Earth is round.

The second of Darwin's discoveries is the way in which so many diverse forms of life were derived from simple common ancestors. Darwin made two observations. The first, the most important, tells us that within every living species there is great *variety*. We humans, for instance, are all different from each other. Dogs are even more different from each other than we are. This variety is a universal characteristic of living beings and is always being renewed. The living constantly change and diversify.

The second observation tells us that in nature only a

minority of living beings successfully reproduce. The majority die before doing so. For all our civilization, this remains true even for human beings. Most fertilized eggs do not result in births.

If we put together this variety with the fact that only a small part of it reproduces, it immediately follows that living beings continue to change, continually experimenting with countless variants, and that only some of those variants – those most able to survive and reproduce in the surroundings they find themselves in – prosper. The others perish. The living beings that we see around us are those that had the characteristics best suited to allow them to prosper.

The understanding of this mechanism has had a huge cultural significance. In the living world many structures, behaviours and forms appear to be accurately designed for life to thrive. Why is this the case? This was a question that had remained open ever since it was raised by Empedocles and Aristotle. Darwin found the full answer.

The answer is that the question was wrong in the first place. It is the wrong way round. It is like asking why there are doors attached to handles. There can be no reason for attaching a door to a handle. But there is good reason for attaching a handle to a door. Living things do not possess adaptable structures for mysterious reasons: it is only those that have adaptable structures that are living in the first place.

The ramifications of this discovery are far-reaching and contribute significantly to our understanding of the deep nature of things. It shows us that the seeming finality of the biological world is only the result of the richness of combinations of things of which the world is made. There is no such thing as intentionality in nature. It isn't design that directs the combination of things, but it is the combination of things that gave rise to intentionality. This has led us to take another step back with respect to the naïve animism of antiquity.

It is still possible to believe in the existence of God the creator of the universe even after this discovery, and there are of course many who do. But the discovery of these simple, basic facts about nature has rendered inconsistent various traditional arguments that purportedly showed the necessity of a divine will to make the world go round. Just as understanding where rain comes from or what causes lightning prompted faith in the existence of Zeus to evaporate, so too the understanding of how life evolved and diversified on Earth has vastly multiplied the number of atheists in the world.

Marie Curie

There is only one scientist who has won two Nobel Prizes for their work in two different sciences, and she is a woman, Maria Skłodowska Curie, better known in her adopted France and throughout the world as Marie Curie. She was awarded a Nobel Prize in physics for the discovery of radio-activity, and quite a few years later a Nobel Prize in chemistry for the discovery and analysis of two new elements: radium and polonium.

She managed to do this, and much else besides, while struggling against adverse and even hostile circumstances. Firstly, because she was a woman in a world where it was normal to assume that women were inferior and should be subordinate to men; secondly, because she was a migrant in a world where immigrants were looked on with suspicion and malice.

People spoke at the time, as they do again now, of the 'clash of civilizations'. The expression 'clash of civilizations' was coined at that time, and the two civilizations that were supposedly in imminent danger of collision were those great allies of today, France and Germany. Maria was Polish, and as such was regarded with disdain and hostility by the French.

She arrived in France relatively late; her family was culti-vated but poor, and she had to work in Poland to support the education of a younger sister. Maria's father was an atheist; her mother a practising Catholic. In France she met Pierre Curie. They shared a passion for science, and for each other.

She was married in a civil ceremony wearing a dark blue out-fit that she subsequently adapted and wore for years as a lab coat. Their laboratory was an old, ill-equipped warehouse which they had converted as best they could. Her collaboration with Pierre was prolific, but in those days it came at a cost for Maria, since everyone took it for granted that it was the man who must have had all the ideas.

When the Nobel Prize was awarded for the discovery of radioactivity, it was initially given only to Pierre. Pierre, knowing that his wife was the greater scientist, convinced the Nobel Committee to divide the award between them – and thus she became the first female laureate. Pierre died soon after this in a road accident, run over by a carriage.

The Nobel Prize allowed Maria to have the funds for research, but it also increased her visibility and focused xenophobic resentment. A relationship with a younger French mathematician, Langevin, unleashes public opinion against them. News of the affair spreads while she is attending a conference, and on returning she finds a hostile crowd outside her home, forcing her to seek refuge with friends.

One of the most impressive chapters in Maria's life takes place during the First World War. She sees how X-rays can have a ground-breaking medical application and constructs the first (mobile) radiography units, which she takes to the front. The number of these units is rapidly increased, and it is estimated that a million soldiers were treated with the benefit of these units during the war, saving thousands of lives.

With the discovery of radioactivity, of its chemical and physical properties and its medical uses, Marie Curie opened the gates to the great science of the twentieth century. Thanks to radioactivity, physicists began to understand the structure of the atom. The difficult life of Marie Curie, her courage, rigour and integrity, are a source of inspiration to everyone, to both women and men of science alike. Her clashes with the bigotry, obtuseness and downright stupidity of patriotism and 'defence against foreigners', her unstintingly serious gaze, the richness of her scientific legacy and the clarity and generosity of her life have remained exemplary. Albert Einstein said of her that she was the only person never to be corrupted by fame.

The Master

When science and religion appear to be on a collision course, it is good to remember the Catholic priest who was also a great scientist, with a capacity to move with rare mastery between science and faith. To the point of getting the better, in quick succession, of both Einstein and the Pope.

His name was Georges Lemaître. An appropriate enough name, since in French Lemaître sounds like 'the master'. And yet it was ill suited to a character as reserved and humble as he was. Having been educated by Jesuits, having taken his vows and distinguished himself in his native Belgium, Lemaître was appointed as a lecturer at the Catholic University of Leuven. He had become fascinated by Einstein's recently published theory of general relativity, and above all by the possibility suggested by Einstein himself of studying the large-scale dynamics of the entire universe: what today is called 'cosmology'.

Einstein had begun this study but had soon realized that his theory predicted that the universe could not be static. The reason for this is easy to understand: the galaxies attract each other, and if they were not moving they would fall one on top of the other, just like balls that cannot stay suspended in mid-air. Einstein did not have the courage to take the prediction of his own theory seriously: the idea of the large-scale movement of the universe just seemed too audacious.

Lemaître sought to change his mind. A ball does not fall if it has been kicked and is flying upwards. The stars might similarly be in flight, distancing themselves from each other, 'kicked' by some tremendous initial impetus. This is the expansion of the universe.

How can we determine whether this is true or not? Lemaître understands that we can do so by observing the light that emanates from distant stars, because the light of something that is receding from us is reddish in colour. He gathers together the small amount of data available on nebulae, the small opalescent discs that appear between stars and which at the time were beginning to be suspected of being very far away indeed, and finds that their light is in fact reddish. Hence this seemed to indicate that the universe is indeed expanding. He publishes these findings in an obscure French scientific journal that no one reads.

A few years later the American astronomer Edwin Hubble studies the nebulae, using the large Mount Palomar Telescope, and shows that they are extremely distant and moving further away from us at great speed. The universe really is expanding! Einstein is obliged to recognize that Lemaître was right.

The young priest extrapolates the consequences of this discovery. If the universe is expanding, it must at first have been extremely small. Lemaître calls this initial state 'primeval atom'. Today we call it the Big Bang. In the years that follow, the idea that the universe emerged from this Big Bang becomes increasingly well known. On 22 November 1951, Pope Pius XII speaks fulsomely about the theory in a public address. The Pope enters into detail about modern science in order to argue that 'the more real science advances, contrary to what was vaunted in the past, the more it reveals the presence of God – almost as if He was waiting patiently behind every door that science opens'. At the core of the papal argument is the Big Bang: 'Creation is thus temporal; therefore there was a Creator; therefore God exists!'

Lemaître is not pleased. In close contact with the Vatican's scientific adviser, he immediately goes out of his way to convince the Pope not to make any further statements of this kind, and to refrain from making references to possible links between cosmology and divine creation. The well-balanced conviction of Lemaître, developed in many of his writings, was that neither science nor religion should attempt to speak of things in areas in which they have no competence. Religion, according to Lemaître, should concern itself with our soul and with salvation, leaving to science the understanding of nature. He was convinced that it was foolish to defend the idea that whoever wrote *Genesis* had even the slightest understanding of cosmology. *Genesis* knows nothing about physics, and physics knows nothing about God. Pope Pius XII was persuaded. He never made any further public reference to his theories.

The idea of using the theory of the Big Bang as an argument in favour of the existence of God reappears in some Protestant contexts in the United States, but the Catholic Church has left it alone. Lemaître, of course, was right: today's discussions have moved on to consider whether the Big Bang could have been a transition from a prior phase in the life of the universe. It is possible that the universe rebounded after a violent contraction. It would have been embarrassing for the Church to have in effect suggested that the Lord, breathing on the waters, had uttered not his primordial 'Fiat lux' but something to the effect of 'Let there be, again, the light that just went out.'

It doesn't fall to everyone to prove Einstein wrong, or to successfully contradict and dissuade a Pope. To have engaged with both individually, and to have convinced both that they had erred on such major issues, is surely something of a result. The 'Master' really did have something to teach.

And yet what Lemaître had to teach us, perhaps the secret of his greatness, is to be found elsewhere.

In 1931 a group of physicists decided to make known the article in which Lemaître had first announced his Big Bang theory. Translated into English, the article was republished in a well-known journal. And it is at this point that a strange kind of detective story begins, with the solution to the mystery only recently revealed by the astrophysicist Mario Livio. In the English translation of 1931, there are a few crucial phrases missing. Precisely those which make completely clear that Lemaître had already deduced the expansion of the universe on the basis of the sparse data available to him at the time, before Hubble. It was as if someone had deliberately erased precisely the phrases that would have cast doubt on the American Hubble's claim to have got there first, and that would have showed it was Lemaître who was the real discoverer.

Who did it? When the alteration of the article came to light, suspicions quickly multiplied. Who was interested in having Hubble take the glory? Was it Hubble himself? The editor of the journal who wanted to avoid alienating the Americans? For years, the missing phrases have provoked arguments and accusations – until the truth emerged from the correspondence between the editor and the author. The removal of the crucial phrases, and the misplaced attribution of the major discovery, was the sole responsibility of . . . Georges Lemaître himself. In a handwritten letter to the editor, he points out that Hubble's data was superior to what was available to himself, and that therefore there was no reason to refer to less precise data that had now been superseded. In other words, Lemaître was uninterested in taking the credit for the discovery. What mattered to him was not personal recognition but establishing the truth.

The man who first saw the Big Bang, the man who knew how to convince both the Pope and Einstein, was curious about nature, uninterested in his own ego. His message seems to me the deepest and clearest that science has managed to articulate: don't take yourselves too seriously; stay humble. Even if your name is Einstein, even if you are the Pope himself. Even if you are 'The Master'.

Which Science is
Closer to Faith?

At a small conference organized by the *Specola Vaticana*, the observatory of the Vatican, a select group of scientists, including a few Nobel Prize winners, was brought together in the beautiful setting of Castel Gandolfo, the summer residence of the Pope on the hills behind Rome, in order to exchange ideas on the theme of 'Black Holes, Gravitational Waves and Singularities in Spacetime'. In the growing, bleak panorama of obscurantism, this small group of scientists represented a point of light, depth and rationality. I remember a previous visit to Castel Gandolfo, several years ago, when George Coyne was the director of the observatory – a profound thinker whose words and writings have left their mark on me. At the time I also met the current director, Guy Consolmagno; in the passion with which he spoke to me about 'his' meteorites I recognized my own love of the universe and its mysteries; my love of science.

The conference was dedicated to Georges Lemaître, the great scientist who is still less well known to the public than he deserves to be. Lemaître was a Catholic priest, deeply interested in the relations between religion and science, a subject on which he wrote pages of great current relevance, and which in my modest and inexpert opinion are illuminating. A good friend of mine, who is himself a priest and a scientist at the Specola, drew to my attention a text on

Georges Lemaître published in the *Commentarii* of the Accademia Pontifica by Paul Dirac. That is to say by one of the two greatest physicists of the twentieth century (the other being Einstein).

Dirac was a man of very few words who was probably affected by a kind of mild autism. He was also a complete atheist. His article about Lemaître, published in 1968, is highly technical. Dirac shows with his usual acumen the significance of Lemaître's contribution to science and gauges its intrinsic merit. It is written in his characteristically dry and factual style. There is, however, a passage that prompted my present thoughts and has given rise to this reflection.

Towards the end of the article, in a somewhat uncharacteristic way, Dirac allows himself to meander into vague speculations about the relation between the cosmos and humanity. Lemaître discovered the fact that the universe evolves. Dirac says that this discovery suggests a great vision: cosmic, biological and social evolution go hand in hand, carrying us to a better and brighter future. It was 1968, and perhaps even the elderly and austere scientist was allowing himself to be influenced by the fevered atmosphere of optimism and anticipated change that could be felt in that remarkable year.

But Dirac cites this consideration in order to then relate a conversation he had with Lemaître on the subject. Moved by the grandeur of the vision that Lemaître had opened up for us, Dirac had told him that cosmology could be 'the branch of science closest to religion'. Perhaps in the sometimes awkward manner of someone on the autistic spectrum, the atheist Dirac wanted to say something kind to the priest Lemaître.

But to Dirac's astonishment, Lemaître disagreed. And after

a brief reflection responded that, in his opinion, cosmology was not the branch of science that was closest to religion. So which one is? asked a perplexed Dirac. Lemaître had a ready answer: psychology.

Lemaître had been at pains to keep cosmology and religion separate from each other. I believe it was thanks to him that the Catholic Church did not fall into the trap that other denominations have found themselves in by attempting to connect the Big Bang with the creation story told in *Genesis*. But the idea that the science closest to religion is psychology, coming from a Catholic priest who has thought deeply about the connection between religion and science, is still a surprising one. Surprising, but to my mind also revealing.

A few months ago there was an article in *Nature* by a representative of the Anglican Church. It was a sincere appeal to put aside the traditional conflict between science and religion and to focus on what they have in common, rather than on their differences. To look for points of convergence instead of conflict is always good advice, and Lemaître's thoughts on the subject seem to me enlightening for precisely this reason. The similarity of the language used by science and religion ('universe', 'creation', 'foundations', 'existence', 'non-existence', 'creator' . . .) seems to me to be simply illusory and therefore essentially misleading. The debate between the two camps is like an argument between two people who cannot hear each other. And they are using the same terms to mean different things.

There are some religions that do not find themselves in conflict with the scientific world; others feel distinctly threatened by scientific thought and take up arms against it. Where does this difference come from? It seems to me that Lemaître put his finger on a significant reason. Religions are complex cultural and social structures that have played a significant role in the evolution of civilization. For a long time, they have identified with the exercise of power and with public affairs, and have offered a complex and global framework for thinking about reality, including about such questions as the origin of the universe. When humanity found better ways of dealing with various of these questions – secular democracy, for instance, with its tolerance and pluralism in the conduct of public affairs, or science, with its interrogation of the world at micro and macro levels – some religions were unsettled by their consequent loss of relevance and entered into conflict with what they saw as pernicious innovations. The aggressive diatribes of Pope Pius XII against religious liberty, against the freedom of the press, against liberty of conscience, are examples that I suppose the current Church must find embarrassing. The siege mentality of the Church, its defence to the bitter end of its central role in public life, which we would like to see consigned to history, is a retrograde battle against science, revealing its incapacity to understand the beneficial and positive growth and evolution of morals. All of this has nourished, and still very much feeds into, the discrediting of the Church with many citizens.

But this does not mean that religion and science must always be in conflict. There are great religions that have no difficulty in accepting the fact that the physical history of the universe is not made more intelligible by reading a religious text or taking on trust what has been handed down to us by tradition. They have no difficulty in accepting the secularism of public life, the plurality of opinions, the genuine tolerance of diversity, and the idea that none of us, in or outside of this or that Church, is the custodian of absolute truths. At their best, the Anglican Church and some forms of Buddhism provide examples of this. They do not try to impose their point of view on those who do not share it, or their behaviour on those operating with a different set of morals; they do not have the presumption of teaching what they do not know themselves. But they do know how to offer admirable examples, and to speak in a convincing manner to the condition of men and women; they know how to bring to bear deep reflection on the nature of humanity, on our choices, our relations, our inner being. Reflections that have a real and deep value for millions of men and women. They know how to offer teaching, transcendence, rituals, cohesion, refuge.

They are religions aware of the fact that their real knowledge concerns our inner life, the meaning that we choose to give to our lives, and not the world around us, not the laws governing public affairs, and not our understanding of the physical universe. These are the religions which know that they have nothing to do with cosmology.

In fact, they are eager to learn from quantum mechanics or from cosmology. The Dalai Lama is one such student, as are the scientists attached to the Specola Vaticana. They are religions, too, that know how to usefully engage in dialogue with the science that is closest to religion: psychology. This is what Lemaître was getting at, with a profound sense of the importance and the current limits of science, but also of the proper importance and limits of religion.

Leopardi and Astronomy

Giacomo Leopardi is, with Dante, Italy's greatest poet. But poetry wasn't his unique interest. His *History of Astronomy: From Its Origins until 1813* is astonishing. It consists of three hundred pages of dense erudition in which he traces the evolution of astronomy from antiquity until his time, the nineteenth century, citing with punctilious detail all of the sources, with a comprehensiveness and an expertise that is perhaps not to be found (and they must forgive me for this) even among my colleagues who are historians of science. And all this was achieved by Leopardi when he had reached the ripe old age of . . . fifteen!

This book by the young poet is one of the clearest testimonies to the intersection between science and literature that is at the heart of the best of Italian culture. The cultural formation of our two greatest poets, Dante and Leopardi, includes extensive and in-depth knowledge of the science of their time; deeply absorbed, properly understood, becoming in the process a genuine source of their poetry.

Leopardi covers the history of astronomy from its obscure beginnings among the Chaldeans right up until the year his text was completed. It is not a treatise on astronomy. He does not go into technical detail where he has no competence to do so. But this makes even more impressive his lucidity when evaluating diverse contributions to the subject, and his ability to extract the pith of all the major results. With an almost uncanny maturity for his age, he manages to combine an endless wealth of bibliographical detail with a clear presentation and synthesis of it. How a boy of fifteen could acquire such erudition and the capacity to digest and explain the subject is something of a wonder. Even at this age, Leopardi was exercising the exceptional intelligence that many who came across him in person remarked upon.

The meaning of the work becomes clearer if we take into consideration the context from which it emerged. Leopardi was living in isolation in Recanati, a small provincial village in a remote part of central Italy. His father, Monaldo, a conservative closely adhering to the authority of the Church, was strongly opposed to the ideas of Copernicus, while Giacomo at the age of fifteen passionately believed in them. For all its erudition, the book is really an act of rebellion.

As in his *Essay on the Popular Errors of the Ancients*, which follows his work on astronomy, science is felt to be a tool for growth, for eliminating the errors of ignorance and bigotry which the young Leopardi perceives all around him. This is the period of what he later called his 'mad and desperate' studies, during which he practically teaches himself Latin, Greek and Hebrew, as well as English, French and Spanish. He is completely absorbed by this.

> Solace and laughter,
> Sweet offspring of our youthful days,
> And you, Love, the brother of youth,
> The bitter sigh of later days,
> I care nothing for. Not knowing why,
> I almost flee from them.
> Virtually solitary and estranged,
> From this my native place
> I let the springtime of my life fly by . . .

His freedom is the extraordinary library that he finds in his own home. There he educates himself, discovers new worlds, dreams. The ideas of the European Enlightenment find their way to him and ignite his mind. He dreams of escaping, beyond 'this hedge'; beyond those 'far mountains'. His young, generous heart rebels against the obtuse, blinkered, dominant thought of the Papal States, and discovers worlds. Astronomy gives him means for this escape from the self, and to journey towards the infinite. To look up at the heavens, to speak with the stars, with 'the vague stars of the Bear', to converse with the moon – 'What are you doing, moon, in the sky? Tell me, what are you doing?/ Silent moon' – in a manner that will always remain the signature of his lyrics.

Perhaps if he had immediately succeeded in escaping from Recanati at his first attempt, propelled by his youthful enthusiasm and the awareness of his own talent, if his father had not discovered and stopped him, clipping his wings, Leopardi's life might have turned out better. But then, perhaps, we would not have had his poetry.

The path is not going to be easy for him. He speaks of feeling deceived by the promising enthusiasms that swept through him at this time. The Truth he pursued with passion unmasks his illusions, and he doesn't have the strength to live by it. His profound sincerity prevents him from lying, from finding refuge in anything that smacks of falsehood, but he is still too tied to the past to accept with any joy the lightness and the freedom achieved. For him, the result is unhappiness; for us, it some of the purest and most movingly beautiful poetry ever written.

The *History of Astronomy* was not published in Leopardi's lifetime. Perhaps he was not wholly convinced by what he had written. It is still a valuable source of information for an academic (I used it when preparing my course on the history of science), but it is not exactly a great read. Leopardi is aware of its shortcomings; the book ends with the words, 'If the present age does not care for my work, at least the sacred shadows of those who contributed to the advancement of the science of the stars might show some gratitude.' It is for himself, in the end, that Leopardi has written this monumental work, just as many of us in our adolescence filled notebooks with no other purpose than to help us grow. Leopardi is seeking nourishment for his soul, and he finds it in what we know about this vast world: in science, in astronomy. 'Man,' Leopardi begins, 'rises above himself by means of it [astronomy], and comes to know the cause of more extraordinary phenomena.' The primary source, as much for his intellectual clarity as a poet as for the enchanted way in which he looks at the world, was the fact that he had made contemporary scientific knowledge deeply his own, beginning with astronomy, the mother of the sciences.

It is difficult, perhaps impossible, to know the lyrics of Leopardi and not to love him – not to feel he is like a brother; to fail to recognize in him one of the most intense and authentic singers of our soul. His awareness of the 'infinite vanity of everything' is one of the most honest afforded by our literature. And yet, at the same time, his own cantos show, not in theory but directly, working upon our deepest emotions, how the ecstatic beauty of the world, the fragrance of *ginestra*, is ultimately sufficient.

In speaking of the meaninglessness of life, his cantos make everything overflow with meaning.

Leopardi feels close to us because he speaks the language of the heart when it is lost, disillusioned, naked before the truth. But he is closer still because, despite his disillusion, the marvellous thing about his cantos is that they give meaning to the beauty of the world; they give significance to everything. In Italy, many of us have loved him, in our difficult and solitary adolescence, and his poetry continues to resonate with us, telling us that life, despite the 'infinite vanity of things', is also an enchantment. As in his most famous line, 'it is sweet to us to be shipwrecked in this sea'.

De rerum natura

In 1417 the Florentine humanist Poggio Bracciolini discovered in a German monastery a copy of the *De rerum natura*, the extraordinary poem by Lucretius which had been forgotten for millennia. Bracciolini could hardly have imagined the influence that the short text in poor condition which he held in his hands would go on to have. The extent of that influence on the Italian and wider European Renaissance, and in fact upon the entire development of the modern world, is reconstructed by Stephen Greenblatt, one of the principal proponents of 'New Historicism' in Anglophone literary criticism, in *The Manuscript: How the Discovery of a Lost Book Changed the History of European Culture*.

A vision of the world which had been almost completely swept away by medieval monotheistic absolutism was re-emerging in a changed Europe. It isn't just the naturalism, rationalism and materialism of Lucretius that are being awakened in Europe. It isn't only a luminous and calm meditation on the beauty of the world and the possibility of serenely accepting death. It is much more than this: it is an articulate and complex conceptual structure for thinking about reality, a new and radically different way of thinking to the medieval one that had dominated for centuries.

The medieval cosmos, so wonderfully depicted by Dante, was a spiritual and hierarchical organization of the universe which mirrored European society; with a cosmos centred on the Earth; the irreducible separation between the Earth and the Heavens; with finalistic and metaphorical explanations of all phenomena; with the fear of God and death; and the idea that it is eternal forms preceding actual things that determines the structure of the world, as well as the idea that the sources of knowledge are located exclusively in the past, in revelation and tradition. There is no trace of any of this in Lucretius. There is no fear of the gods, no intentionality or causes in the world, no cosmic hierarchy and no distinction between the Earth and the Heavens. There *is* a profound love for nature, a calm immersion in it, a recognition that we are part of it ourselves; that men, women, animals, plants and clouds are organic parts of a wonderful whole, a tissue without hierarchy. There is a profound universalism, and there is the aspiration to think of the world in simple terms. To be able to investigate and eventually understand the secrets of the physical world. To be able to learn more than our fathers and forefathers knew. And in Lucretius, remarkably, there are the conceptual tools that Galileo, Kepler and Newton will build upon: the idea of free and rectilinear movement in space; the idea of elementary bodies – the atoms – which through their combinations weave the complexity of reality; the idea of space as a container of the world. Above all, there is a measured and passionate defence of the notion that existence can be serene even though it is limited; that we should not fear death, precisely because there is nothing beyond death. And that we should not fear God, because, even if He existed, He would be too busy with more momentous things to be distracted by us, such irrelevant grains in a boundless cosmos. The echo of this mental universe that was resurrected by the rediscovery of Lucretius, reverberates directly in the pages of authors ranging from Galileo to Kepler, from Bacon to Machiavelli, from Montaigne, who in his essays cites Lucretius more than a hundred times, down to Newton, Dalton, Spinoza, Darwin and even Einstein, who wrote beautifully on Lucretius: 'For anyone who is not completely submerged in the spirit of our age . . . on him will Lucretius's poem work its magic.'

A recent book by Piergiorgio Odifreddi, *How Things Are: My Lucretius, My Venus*, offers a readable prose version of Lucretius's poem, with an extensive commentary that illustrates with examples from contemporary science the rational legibility of the world, in the manner of Lucretius himself. Odifreddi reprises in contemporary terms the great themes that animate Lucretius's vision: the love of nature as the sole artificer of everything, the faith in the power of reason that allows us, step by step, to understand it and which dispels irrational fears caused by death and by religion. He restores to us, revived and luminous, this great Lucretius who bridged the gap between Greek atomism and ourselves, and whose work constitutes for that reason one of the deepest and most vital cultural roots of the modern world.

Alongside Odifreddi's text, I reread Vittorio Enzo Alfieri's *Lucretius*, first published in 1929. Here, to my surprise, I found a reading of Lucretius's poem that is opposite to Odifreddi's. Whereas Odifreddi's Lucretius reflects the serenity of reason, Alfieri's is perceived to be some kind of tortured romantic. Alfieri is blind to the luminous ideas that Odifreddi clarifies, to the conceptual clarity and to the immense intelligence of Lucretius's reading of the world. He sees something else instead. He hears the music of the work and responds to the marvellous poetry of nature and the passionate soul of Lucretius – his subtle sensibility.

Alfieri holds our hand and guides us through verse after verse, pointing out the dazzling beauty of the poem, showing us its secret rhythm, its musical quality, now immense, now intimate, and deciphers in its texture the crepitations of the author's heart.

In Lucretius's passion for reason, Alfieri deciphers a kind of desperation. The song of Lucretius shows the foolishness of men, the uselessness of life, the absurdity of consoling illusions. Lucretius dwells at length – at great length – on death, ending with the realistic and raw description of the horror caused in Athens by the plague, in 'painful lines by the poet of the serene life'. For Alfieri, Lucretius's passionate declaration of faith in the serenity of life is like the yearning, almost wishful, thinking of someone who has suffered greatly. He reprises the highly dubious tradition according to which Lucretius took his own life while crazed by a love potion, and reads the suicide of the poet as an act of heroic resistance in the name of reason, choosing an ultimate dignity in order not to give in to that black sea that Conrad glimpsed behind the surface of reality.

Alfieri's Lucretius is a kind of romantic titan, motivated by heroic rebellion, on behalf of man and against the foolishness of religion and the illusions of love, who wants to offer to himself and to the rest of us a path to knowledge and serenity – but whose project collapses because nature for him is not so much a caring mother as a wicked stepmother, and because the passions of the heart are much stronger than serenity of thought.

Which interpretation is right? The unillusioned serenity traced by Odifreddi, who laughs at the gods; or the turbid romanticism of Alfieri, who trembles with emotion when reading the poetry of Lucretius?

Perhaps they are both right. Lucretius is deep and wide enough to contain all this, and much more besides.

But what I believe really interests us, and the teenagers who are drawn to his work, is not who Lucretius was: it is life itself. How far can we go in understanding our reason? Can it save us from the monsters that dwell in us? Or should we renounce lucidity in order to find consolation? Can we be enchanted by his understanding of reality and at the same time transported by his poetry? Is it possible to seek the light of thought without becoming myopic with regard to the infinite complexity of what is happening in front of us? Is nature a mother, or a step-mother? Does the lucidity of naturalism lead to the desperation of Leopardi? Or to the serenity Lucretius invites us to share? Does understanding make us free?

At the end of the fourth book of *De rerum natura* there is one of the most desacralizing, wildest descriptions of love ever written. Love is returned to its most brutal physical source:

> When at last with limbs intertwined
> They enjoy the flower of their youth
> And the body anticipates already the pleasure to come
> And they have reached the point at which Venus
> Sows seed in the field of womankind,
> They tighten against each other avidly,
> Mingling the saliva in their mouths
> And, pressing lips against teeth,
> Each inhales the breath of the other –
> In vain, since nothing can detach from her body,
> He cannot penetrate it with his entire body,
> Losing himself in that body.

It's a passage that leaves you breathless.

And yet Alfieri himself, flabbergasted, recognizes that rarely has anyone come as close as this to the essence of love, to its turbulence and hunger. At the very moment in which Lucretius strips it most, the closer he gets to capturing its wordless essence. And this, for Alfieri, is the irresistible love that killed Lucretius.

And yet it is the same voluptuousness that opens the poem and suffuses it with joy. To paraphrase, it begins like this: O Venus, o voluptuousness, you are the spring, the sun, the desire, the fecundity of animals and earth; before you, winter, sadness, death flee . . . for you, the level oceans laugh, the calmed sky shines with infinite light . . .

Lucretius confronts us with reality, in all its complexity. The desperate melancholy of life, the luminous joy, an endless vision of the cosmos, moving lyricism, the contemplation and understanding of nature, the constant desire for knowledge. Why is a poem as vital as this not read in Italian schools – or schools everywhere, for that matter? Perhaps it would speak especially to teenagers about what is happening inside them.

> We move in circles, always at the same point [. . .] our
>   appetite for life
> is voracious, our thirst for life insatiable. (III, 1080–84)

All human life really is here: atoms and the cosmos, physical fields and the invisible, ambition, infidelity, boredom, religion, fear, death and serenity before the tragic questions of men faced with death, and a vortex of cosmic life – from the dance of a speck of dust in a beam of sunlight, to the dissolution of the world in the far-distant future, eons away. Why not give to young students both interpretations of Lucretius, and leave it to them and the generation that will follow us to try to resolve what we have not been able to decide?

Do Flying Donkeys Exist?
David Lewis Says Yes

'Who is David Lewis?'

'He is one of the greatest philosophers of the century.'

'Goodness. And what does he believe?'

'That all possible worlds actually exist.'

'But what can that mean, it makes no sense: do you believe it?'

'No.'

This is a rather surreal conversation, for how can you claim that somebody is the greatest philosopher in the world and then immediately add that his main thesis is not even credible? And yet I have had this conversation, almost verbatim, a surprising number of times with a surprising number of eminent philosophers of various nationalities. In the somewhat rarefied atmosphere of analytic philosophy, the American philosopher with strong links to Australia, who died just over a decade ago, is today recognized by many of his colleagues as one of the greatest, if not the greatest, of all contemporary philosophers – even if his most well-known thesis, the concrete existence of many worlds, leaves many scratching their heads.

Lewis was an appealing character. In his lectures he was prone to talk about science-fiction movies and time travel. His philosophical articles are crowded with flying donkeys, cats that lose all their fur on sofas, and so on. He had a bushy beard, and an estranged air about him. He exuded something unconventional and slightly unhinged. He loved Australia, where he spent many months each year. He wrote dozens of articles on the widest range of subjects, all of interest to analytical philosophy, and a few books, the most notable of which is *On the Plurality of Worlds*. As its title suggests, it focuses his main thesis: that all possible worlds exist. Including those in which donkeys fly.

But wait a minute, you will rightly say, there are no flying donkeys in this world. And Lewis agrees, there are no donkeys that can fly *in this world*. But in other worlds there are. There are many worlds in which donkeys fly. All of these worlds are really there. They exist. We do not see flying donkeys because we live in a world where they do not fly. Just as from my window I can see the sea of Marseille and not the Colosseum, because I am in Marseille and not Rome. Not because Marseille exists and Rome does not. Just as Rome exists even when I am not there, so other worlds exist even though we are not in them. And which of these other worlds exists, exactly? They all do, replies Lewis, behind a disarming smile.

Slightly perplexed by hearing about these ideas, last year I decided to try reading Lewis, even though I am not a philosopher and don't have all the necessary conceptual tools at my disposal.

The first thing I read was an article about time travel. Given that my own subject is physics, and that I am particularly interested in the nature of space and time, I imagined that I would be in quite a strong position to tackle philosophy in a field with which I am familiar. Besides, I've always found that what I usually read on the impossibility of time travel is confused and messy, and I expected more of the same. To tell the truth,

I began reading the article fully expecting to catch the supposedly great philosopher making some basic, elementary mistake. Instead I soon found myself open-mouthed. Lewis discusses the possibility of travelling in time with complete clarity. His article is utterly, unequivocally clear. He has brought perfect order to the question. All of the foolish assertions about how, if we went back in time, we could end up by killing our grandfathers are swept aside with lucid simplicity. I began to see why so many people are dazzled by David Lewis.

And so I immersed myself in a collection of his articles. A few of the more technically philosophical ones bored me, frankly, and I failed to understand them. But many struck me as exceptionally brilliant. What exactly, asks Lewis, is a thing, an entity? A cat that is lying on a sofa, for instance. Does it include the hair that has left the cat and adhered to the sofa? Where exactly does the cat end and its context begin? And so on and so forth, zigzagging between technical problems of modal logic and those questions we used to discuss as teenagers and never found answers for. For every question, Lewis has a convincing answer, even though he always presents them with a smile on his lips. He finds solutions precisely where there seemed to be none, with dazzling intelligence.

Armed with this experience, I felt ready to tackle his major work, *The Plurality of Worlds*. Now let him try to convince me, I found myself muttering, that flying donkeys exist . . .

I have to admit defeat, to confess that he really did manage to persuade me. I will not even attempt to summarize how. I am not Lewis. If you are at all intrigued, do read his book. I still have a few doubts. I still ask myself, for instance, whether Lewis does not merely change the meaning of certain words, using 'to exist' where others would use 'to be possible', and calling 'of this world' what others would call 'existent'. But he has undoubtedly changed my ideas about what 'to exist' means. He has helped to free me, I think, from prejudices that are still attached to this notoriously slippery verb. Does a puppet exist whose nose grows when he tells lies? Yes, of course it does, it's Pinocchio. So Pinocchio exists? No, he doesn't exist! But you just said that he exists . . .

At the very least, Lewis has completely convinced me that to speak of possible worlds as if they were real is a very effective tool for achieving clarity in all questions involving modality, which is to say those arguments pertaining to possibility or necessity.

Thanks to Lewis and his colleagues, analytical philosophy has returned to engaging weightily with metaphysics – terrain that for so long it had kept a safe distance from. The lesson of logical positivism, which had insisted upon affirming that we should speak only of those things that could be defined in a sufficiently clear manner – and above all the one provided by Wittgenstein, who showed how many apparently profound problems are nothing but the result of clumsy and inexact uses of language – had left a deep mark on this vast area of philosophy, with the result that questions revolving around existence and non-existence had become habitually regarded with scepticism in these quarters. Even today the word 'metaphysics' has the capacity to prompt a collective raising of eyebrows in some philosophy departments around the world. Lewis himself is sometimes subject to this kind of suspicion. Yet there is undoubtedly a part of analytical philosophy that has found a way, with the clarity of thought that characterizes it, and with its own tools, of returning to dealing with questions such as what exists or does not exist. Lewis has contributed significantly to bringing metaphysics back to the centre of discussion.

We are Natural Creatures in a Natural World

Naturalism without Mirrors is a complex book in which one of the most brilliant contemporary philosophers, Huw Price, the Bertrand Russell Professor of Philosophy at Cambridge, discusses a version of what it would be no exaggeration to call the dominant philosophy of our time: naturalism. It is a version that responds implicitly to many in-house anti-naturalistic positions.

Naturalism, writes Federico Laudisa in a recent volume of the same name, 'has become a general framework of reference for many philosophical questions at the centre of debates during the last half century'.

Like all major tendencies of thought, naturalism does not have a precise definition, and can be conjugated in a variety of ways. It can perhaps be characterized as the philosophical outlook that believes all existing facts can be investigated by the natural sciences, and that as human beings we belong to nature: we are not distinct entities which are separate from it. You are not a naturalist if you assume there are transcendent realities that can only be known in a way that is not subject to scientific enquiry. You are not a naturalist if you believe in the existence of two realities: one nature that may be studied by science, and another that is impermeable to it.

Naturalism emerges in classical Greek thought, unfolding for instance in the work of Democritus, and is reborn, after a prolonged silence, in the Italian Renaissance, before being reinforced in the triumphs of modern science. It gathers strength in the nineteenth century, and today permeates a huge amount of world culture. Theses that are very markedly naturalistic were championed by Willard Quine, one of the major philosophers of the twentieth century. One of his best known and most extreme propositions in this regard relates to the 'naturalization of epistemology': the effort to redirect to the natural sciences even questions about the very nature of knowledge.

But there are certainly also intellectuals who maintain a distance from naturalism. In his book on naturalism, for instance, Federico Laudisa feels compelled to point out that he 'does not share the enthusiasm for naturalism shown by my colleagues'. Laudisa scolds naturalism above all for not being able to account for the normative (and aesthetic) aspects of thought. More emphatically, Maurizio Ferraris distinguishes 'natural' realities such as mountains, trees and stars, from 'social' ones such as contracts, values and marriages, which are indeed realities but are socially constructed. Though coming from very different traditions of thought, both Laudisa and Ferraris see the limitation of naturalism arising where human thought begins.

This is precisely the problem that Huw Price takes as his point of departure. Price calls it the 'problem of placement', and formulates it as the question of where to 'place' in the world of natural sciences things such as moral values, beauty, consciousness, truth, numbers, hypothetical worlds, laws, and so on: all those things that seem least compatible, for instance, with the world described by physics.

Price's answer is given in two parts. The first is the observation that our language and our thought are not necessarily representations of something external. This observation is at the heart of Wittgenstein's later writing: contrary to what is assumed by the most widespread theory of language (which can be traced to Gottlob Frege, the father of modern logic), our language does a great deal more than designate objects and properties of objects. If I look at the sunset and exclaim 'How marvellous!' to my partner, who is sitting next to me, I am not designating an entity, the marvellous that is out there, in the vicinity of the setting sun. I am expressing the effect of the sunset upon me; I am strengthening the closeness with my partner that comes from being there, enjoying it together, or I am attempting to show something of my inner life. Or perhaps any number of other messages, none of them having anything to do with an external object 'marvellous'. If I say, 'Come here!', I am not designating anything. To interpret language as something that necessarily 'refers' to something external is to create false metaphysical problems. To interpret our sophisticated and complex linguistic activities as affirmations of an external reality is a basic error that, according to Price, generates the false problem of 'placement'.

The second stage of Price's answer involves a subtle slippage of the central idea of naturalism: to accentuate the fact that, as human beings, we are part of nature, and therefore can be studied by the natural sciences. Price calls this 'naturalism of the subject'. The alternative is not between understanding moral values, beauty, knowledge, consciousness, the notion of truth, numbers, hypothetical worlds and so on as a metaphysical furnishing of the world, or instead declaring all of them 'illusory'. There is another possibility: to understand them as aspects of our own behaviour as complex natural beings in a complex natural world.

This does not preclude the possibility of studying them in

an autonomous way: a mathematician studies numbers, a philosopher of ethics studies moral values, and so on. Law, aesthetics, morality, logic, psychology . . . these are independent disciplines. But the presuppositions of these disciplines and the reality with which they are engaged do not contradict naturalism, because they can be reintegrated into and are compatible with the general coherence of the natural world, just as chemistry is compatible with physics: our thought and our inner lives are real phenomena generated by us, natural creatures in a natural world. Many of the liveliest fields of contemporary science are currently engaged in fleshing out this intuition: neuroscience, cognitive science, ethnology, anthropology, linguistics . . . A seemingly endless literature is growing, dedicated to understanding ourselves in natural terms. There is an enormous amount that we still don't understand – because, as always, what we don't know is vastly greater than what we know. But we are learning.

Perhaps in a curious way, transporting ourselves back to our natural reality, which for Price has its roots in pragmatism and in a respect for what we have learned about reality thanks to scientific rationalism, ends up bringing us closer to the intuitions of Nietzsche, which along a different route have led to the excesses of postmodernism: before being a rational animal, man is a vital animal – 'It is our needs that interpret the world . . . Every instinct has its thirst for dominion.' True, but our reason also emerges from this magma, and emerges as our most effective weapon.

Price's book argues with strength and rigour for a humble and complete naturalism: we are natural creatures in a natural world, and these terms give us the best conceptual framework for understanding both ourselves and the world.

We are part of this tremendous and incredibly rich nature

about which we still understand little, albeit enough to know that it is sufficiently complex to have given rise to all that we are, including our ethics, our capacity for knowledge, our sense of beauty and our ability to experience emotions. Outside of this there is nothing.

For a theoretical physicist such as myself, for an astronomer accustomed to thinking about the endless expanse of more than 100 billion galaxies, each one consisting of more than 100 billion stars, each one with its garland of planets, on one of which we dwell for a brief and fugitive moment, like specks of infinitesimal dust lost in the endlessness of the cosmos, this seems no more than obvious. Every anthropocentrism pales into insignificance in the face of this immensity. This is naturalism.

Emptiness is Empty:
Nāgārjuna

We rarely come across a book with the capacity to influence our way of thinking. Even more rarely when it happens to be a book we knew nothing about. But this is what happened to me recently.

I am not talking about an obscure text. On the contrary, it is a very famous one, discussed for centuries by generations of students, venerated even. I had not even heard of it, and I suspect that many Westerners are as ignorant of its existence as I was. The author's name is Nāgārjuna.

It is a short, dry, philosophical text written eighteen centuries ago in India. It has become a classic reference work of Buddhist philosophy. Its title is one of those seemingly endless Indian words, *Mūlamadhyamakakārikā*, which has been translated in various ways, including 'The Fundamental Verses of the Middle Way'. I read it in the English translation by the philosopher Jay Garfield, accompanied by an excellent commentary that helps in coming to terms with its language. Garfield has a deep knowledge of Eastern thought, but his philosophical basis is in the Anglo-Saxon analytic tradition and he presents the ideas of Nāgārjuna with the lucidity and concreteness typical of this school, connecting them with Western philosophy.

I didn't come across this book by chance. Various people had asked me, 'Have you read Nāgārjuna?', often after a discussion about quantum mechanics or some other fundamental aspect of physics. Personally, I have never had much patience with attempts to link modern science and ancient oriental thought; they always seemed forced, and reductive with regards to both. But having heard my latest 'Have you read Nāgārjuna?', I decided to go ahead with what turned out to be for me quite a discovery.

Nāgārjuna's thought is based around the idea that nothing has existence in itself. Everything exists only through dependence on something else, in relation to something else. The term that Nāgārjuna uses to describe this lack of essence per se is 'emptiness' (*śūnyatā*): things are 'empty' in the sense that they do not have autonomous reality; they exist only thanks to, as a function of, with respect to, and in the perspective of something else.

If I look at a cloudy sky – to take a rather basic example – I can see a dragon and a castle. Do the dragon and the castle really exist there, in the sky? Obviously not: they grow out of an encounter between the appearance of the clouds and the sensation and thoughts in my mind; on their own, they are empty entities, they do not exist. So far, so easy. But Nāgārjuna also suggests that the clouds, the sky, our sensations, my thoughts and indeed the very head in which those thoughts are taking place are actually nothing but things which emerge from the encounter with other things: on their own, they are empty entities.

Is it me who sees a star? Do I exist? No, I am no exception to the rule. Who is it who sees the star then? No one, says Nāgārjuna. Seeing the star is a component of that whole, that set of interrelations, which I conventionally call being myself. He also writes that 'What language expresses does not exist. The circle of thoughts does not exist' (XVIII, 7). There is no ultimate or mysterious essence to understand that is the true essence of our being. 'I' is nothing other than the vast and interconnected set of phenomena that constitute it, each one dependent on something else.

Centuries of Western concentration on the subject seem to vanish like morning mist.

Like much philosophy and science, Nāgārjuna distinguishes between two levels. On the one hand, apparent, conventional reality with its illusory and perspectival aspects; on the other, ultimate reality. But he takes this distinction in a surprising direction: the ultimate reality, the essence, is absence, is vacuity. It is not there.

Every metaphysical system looks for a primary substance, an essence upon which everything else must depend: this point of departure may be matter, God, spirit, Platonic forms, the subject, the elementary moments of consciousness, energy, experience, language, hermeneutic circles, or what have you. Nāgārjuna suggests that the ultimate substance ... does not exist.

There are ideas more or less similar to this in Western philosophy, extending from Heraclitus to the contemporary metaphysics of relations ... But what Nāgārjuna is proposing is a more radically relational perspective. The illusoriness of the world, its samsāra, is a general theme in Buddhism; to fully realize it is to reach nirvana, or liberation and beatitude. But for Nāgārjuna, samsāra and nirvana are one and the same thing: they are both empty; both non-existent.

So is emptiness the only reality? Is this the ultimate reality? No, writes Nāgārjuna, every perspective exists only by depending on another, it is never the 'ultimate' reality, and this includes his own perspective: emptiness is also devoid of essence; conventionally so. No metaphysics survives. Emptiness is empty.

Please don't take literally my clumsy attempt to summarize Nāgārjuna: I have certainly not nailed him down. But for my part I have found this perspective surprisingly efficacious, and I keep thinking about it.

In the first place because it helps to give a shape to attempts to think in a coherent way about quantum mechanics, where objects seem to mysteriously exist only when influencing other objects. Nāgārjuna obviously knows nothing about quanta, but nothing prevents his philosophy from providing useful tools for imposing a degree of order on modern discoveries. Quantum mechanics cannot be squared with a naïve realism, still less with any kind of idealism. So how should we think of it? Nāgārjuna provides a potential model: we can think of interdependence without autonomous essence. In fact, true interdependence – and this is his key argument – requires that we forget autonomous essence altogether.

Modern physics swarms with relational notions, not just with regard to quanta: the speed of an object does not exist in itself, it only exists in relation to another object; a field in itself is not electric or magnetic, it is so only in relation to something else; and so on. The long search for the 'ultimate substance' in physics foundered on the relational complexity of the quantum theory of fields and of general relativity . . . Perhaps an ancient Indian thinker can provide us with some conceptual tools to extricate ourselves a little further. It is always from others that we learn, from others different from ourselves; and, despite millennia of unbroken dialogue, the East and the West still have a great deal to say to each other. As in all the best marriages.

But what's fascinating about his thinking goes beyond

the problems of modern physics. It seems to resonate with the best of much Western philosophy, both classical and modern. But it does not fall into the traps that so much philosophy ends up in, by postulating premises that always turn out in the long run to be unsatisfactory. He speaks about reality and of its complexity, screened off from the conceptual trap of wanting to discover its foundation. It's a language close to contemporary anti-foundationalism. It is not extravagant metaphysics: it is simple sobriety. And it fosters an ethical attitude that is deeply comforting: to understand that we do not exist is something which may free us from attachments and from suffering; it is precisely on account of life's impermanence, the absence from it of every absolute, that life has meaning.

This is Nāgārjuna as filtered by Garfield. There are other, different interpretations of his text available. After all, it has been written about for centuries. The multiplicity of possible readings is not a sign of the book's weakness. On the contrary, it is testimony to the vitality and eloquence of this extraordinary ancient text. What really interests us is not what the prior of a monastery in India effectively believed almost two thousand years ago. That, so to speak, is his business. It is rather the strength of the ideas that emanate today from between the lines written by him, and how these lines, intersecting with our culture and our knowledge, can open up spaces for new thought. Because this is the nature of culture: an endless dialogue that enriches us by continuing to feed on experiences, knowledge and, above all, exchanges.

Mein Kampf

The newspaper *Il Giornale* has offered on the newsstands a new edition of Hitler's *Mein Kampf*. There are sound enough reasons for feeling offended or even disgusted by this decision. And yet I found myself in agreement with the newspaper's editor when, perhaps a little clumsily, he sought to defend the controversial act by saying that in order to fight evil you need to know it and to understand it.

For my part, I read *Mein Kampf* some time ago, and I did learn something from it. It taught me a few things that I did not expect to find. I'll try to summarize them here.

Nazism, of course, was a ferocious unleashing of aggression. From the Night of the Long Knives to the desperate defence of Berlin, it rode a wave of extreme violence. The immediate ideological justification for the outbreak of brutality and violence was the self-styled racial and cultural superiority of the Germanic people, the exaltation of force, a reading of the world in terms of conflict rather than collaboration, and contempt for the 'weak'. This is what I thought I would find in *Mein Kampf* before actually reading it.

But Hitler's book turned out to be something of a surprise, clearly showing as it does what was the real source of all this. Namely, *fear*.

For me this came as a revelation which allowed me to grasp something about the mindset of the political right that I had always struggled to understand. A main source of the emotions that give power to the right, and above all to the far right, is not the feeling of being strong. It is, on the contrary, the fear of being weak.

This fear is explicit in *Mein Kampf*; this feeling of inferiority, this sense of being surrounded by imminent danger. The reason behind the need to dominate others derives from a terror of being dominated by them. The reason for preferring combat to collaboration is that we fear the strength of others. The reason why we close ourselves into an identity, a group, a *Volk*, is to create a gang stronger than the other gangs in a relentlessly dog-eat-dog world. Hitler depicts a savage world in which the enemy is everywhere, danger is everywhere and the only desperate hope of avoiding succumbing to it is to band together into a group and prevail.

The result of this fear was the devastation of Europe and the loss of 70 million lives worldwide.

What can we learn from this? I think it teaches us that in order to avoid catastrophes we do not need to defend ourselves against others: we need to fight against our fear of them.

This is what is so devastating. It is this reciprocal fear that inclines us to see others as less human, and that opens the way to an inferno. A Germany offended and humiliated by the outcome of the First World War and terrified by the power of France and Russia was a Germany primed for auto-destruction; the Germany that, having learned its lesson and reconstructed itself at the centre of European collaboration and resistance to war, is a Germany that has truly flourished.

Those who feel weak are afraid, wary of others, they defend themselves and cluster in their supposed identity.

Those who are strong are not afraid, do not seek conflict but collaborate instead, and contribute to building a better world, for themselves and others. If someone tells you that you should be afraid, it is because they are weak. I believe there are few books as revealing of this intimate logic of violence as *Mein Kampf.*

Black Holes I:
The Fatal Attraction of Stars

Ninety-nine years ago, as Europe threw itself sanguinely towards a catastrophic mutual massacre, a thirty-six-year-old Albert Einstein was sending to a scientific journal the article containing the final equations of general relativity. He could hardly have imagined how many and what kind of extraordinary unknown phenomena those equations would reveal.

The equations were complicated, and Einstein did not expect to be able to find exact solutions. And yet just a few weeks later, in January 1916, he received a letter from a lieutenant of the German artillery, who wrote: 'As you will see, the war has been sufficiently accommodating to allow, despite the machine-gun fire, an excursion into the territory of your ideas.' The letter was from Karl Schwarzschild, announcing that he had found an exact solution for Einstein's equations. Four months later, Karl Schwarzschild was dead from an illness developed on the Russian front.

The solution he proposed describes the space surrounding a spherical mass such as the Earth or a star. If the mass is sufficiently extended, it exercises an attraction that is exactly the force of gravity described by Newton three centuries previously, and that we all studied at school. If the mass is concentrated, the force described by the equations of Einstein is more intense than Newton's force, and one of its effects is to slow down clocks. But there is something strange in the solution found by Schwarzschild: if the mass is *extremely* concentrated, this solution predicts a spherical surface where all clocks would stop. Where time would stop passing.

What does this mean?

Einstein makes one of his numerous errors by maintaining that this surface, today known as the Schwarzschild surface, or horizon, could never be reached. He writes an article claiming that there could be no such thing as the objects described by the Schwarzschild solution. The article is wrong. Other theorists join in, and a lot of confusion follows. To understand what actually happens on the Schwarzschild surface we have to wait until the sixties, when mathematicians and physicists begin to untangle the threads and understand that the surface is not an impassable limit. In fact, it can be crossed without difficulty.

It is, instead, the limit of the region where gravity is so strong that nothing, not even light, can escape.

John Wheeler, with his gift for words, finds an appropriate name for this phenomenon: black hole. A black hole is a region where there is a mass that is so compact, so collapsed in on itself, that nothing can escape from its tremendous gravitational pull, not even light. A ray of light on the Schwarzschild surface remains stuck there, without moving, without being able to escape, frozen. Nothing escapes from a black hole; everything can enter into it.

The matter seemed more academic than scientific, because for this 'Schwarzschild surface' to exist requires an incredible degree of compression. The entire mass of our planet, for example, before it could become a black hole would have to be contained inside a marble with a diameter of just one centimetre.

Surely entities as compressed as this could not actually

exist in the universe. Surely it was absurd to expect to squash the Earth into something smaller than a ping-pong ball! Or so it seemed at the time.

Still, when I was studying general relativity at university in the late seventies, my textbook chapter on black holes claimed that they were nothing more than a mathematical curiosity, and that 'There is nothing like them in our real world.'

It was wrong, as is frequently the case with textbooks.

Already in 1972, an extremely compact and dark object in the Cygnus constellation had aroused the curiosity of astronomers. It would become known as Cygnus X-1. There is another star rotating around it at great speed. A black hole, John Wheeler will write, is like a man dressed in black who waltzes in a barely lit room with a woman dressed in white. We know that it is there only because we can see a bright star whirling around it.

The astronomers concentrated their efforts on Cygnus X-1. They managed to observe the light emanating from the matter that ignites as it spirals around it, drawing ever closer before disappearing, swallowed by the void. Soon afterwards, other, similar objects were identified and studied. All alternative explanations of the phenomenon were gradually eliminated, until the conclusion was inevitable: the heavens are full of black holes. Today, it is estimated that in our galaxy alone there are tens of millions of black holes similar to Cygnus X-1.

But there is more. Ever since the early thirties, it was known that transatlantic communications were interfered with by a strange source of radio waves. In 1974 scientists realized that the source of these waves was beyond the Earth, and that they are emitted in the Sagittarius constellation where the centre of our galaxy is. The observations concentrated on this source, called Sagittarius A,* and very gradually something astonishing became apparent: at the centre of our galaxy there is an immense black hole. Its mass is millions of times greater than that of the sun. There are numerous stars rotating around it. Every so often, one of these stars gets too close to this monstrous galactic Polyphemus and is swallowed like a small fish by a whale.

Today, astronomers are putting in place a network of huge radio antennae, extending from the Arctic to the Antarctic by way of the Rocky Mountains and the Andes, which should be able to 'see' the boiling-hot region surrounding the monster, where stars flock uncontrollably together with dust and detritus of every kind, forming a furious vortex in infernal tumult before plummeting into the black well.*

Similarly colossal black holes have been observed at the centre of almost all known galaxies. Some of these are voracious, ceaselessly devouring enormous quantities of stars and interstellar gas. The matter that plummets into them boils violently, reaching temperatures of millions of degrees, producing gigantic rays of energy that light up intergalactic space.

The most violent events that we can observe in the universe, such as the intense and mysterious signals that in the past were known as quasars, are produced by these titans, sometimes as luminous as an entire galaxy of 100 billion stars. Can you imagine a galactic storm being unleashed by a monster one billion times greater than the sun?

* Five years after the writing of this article, the network of radio antennae did indeed succeed in capturing the first image of a galactic black hole. The image has since become famous all over the world.

Black Holes II: The Heat of Nothingness

Stephen Hawking's major scientific contribution concerns the nature of black holes. He has demonstrated that they are hot.

I am not referring to the matter that becomes red hot by spinning and amassing together as it falls towards the black hole, making it visible in the heavens. No: Hawking has shown that even a calm black hole into which nothing is falling is still hot. Black holes are naturally hot.

No one has observed this heat yet. It is too weak to be picked up by any telescope, and in the black holes that we see it is usually overlayered by the tempestuous heat of the matter that is continually falling into it. Hawking's prediction is theoretical at present: it lacks experimental confirmation. But his calculation has been repeated in many different ways, and the result is always the same. The result is judged by the scientific community to be persuasive. A black hole therefore is in all likelihood not so black after all. It is a moderate source of heat. If it was isolated in the middle of a starless sky, it would not seem black but resemble instead a small sphere emitting a pallid light.

This outcome surprised everyone. To be hot means to emit heat. But we thought we had understood that a black hole is a place from which nothing can escape – so how can heat emanate from it?

The key to Hawking's calculation is that it involves

quantum mechanics. Whereas the prediction that a black hole permits only an entry but never an exit is a prediction solely dependent on Einstein's theory, general relativity – and this is an incomplete theory that overlooks quantum phenomena. Hawking's calculation improves our understanding of a phenomenon that Einstein's theory describes only up to a certain point and shows that something – a dim heat – escapes from black holes.

The heat of black holes involves both general relativity, namely the theory that describes the black hole itself, and quantum theory. Currently there is not yet consensus on a complete theory combining general relativity and quantum mechanics, and the heat of black holes is an indication of how to look for this combination. It is a theoretical benchmark for all attempts to solve the problem of combining the two great physics theories of the twentieth century. Black holes are not just amazing real objects in the heavens. They are also a laboratory for theoretically testing our ideas about space, time and quanta.

A cup of tea is hot because its molecules are in a very agitated state. The heat is the rapid movement of the molecules. But the surface of a black hole is not a concrete surface made of matter, like the surface of a ball or the surface of a cup of tea. It is merely a place of no return, where the force of gravity becomes incredibly strong. It is not a material surface composed of molecules. So, to what can we attribute the turmoil on the surface of a black hole, generating heat, if there is nothing there?

One possible answer is that it is elementary quanta of space that generate this heat. The heat of black holes foreseen by Hawking's calculation could be the clue that reveals the existence of these 'molecules of space'. The immensely powerful force of gravity acts on the surface of the black hole like a giant amplifier, revealing the infinitesimal trembling of the elementary grains of space. The heat of black holes is not the heat of some object or other: it is the heat of empty space itself, magnified by gravity. It is the elementary heat of nothingness.

Something puzzling follows from this reasoning: when we seek to combine the theory of gravity with quantum mechanics, it seems that it is not possible to do so without talking about heat. But why is this? Heat can be interpreted as lost information: to say that an object is hot is to say that its molecules are moving around a lot, but randomly, in a way that we cannot precisely reconstruct on the basis of the macroscopic behaviour of the object. If I burn a letter in a fireplace, a super-skilful investigator can in principle trace the text of the letter in the ashes or in the light emitted by the fire; but whatever falls into a black hole is lost for ever to those who are outside of it: if I throw a letter into a black hole, I will never know what was written on it. Black holes destroy all information. Where does it go?

Like a Gordian knot that symbolically closes access to Asia, a black hole is a mysterious object where all the marvels we have recently discovered about the world can be found linked together: time that slows, almost to the point of standing still; elementary quanta of space; lost information. A place in the universe where everything can enter and nothing ever leave, for all eternity . . . It causes a sense of unease. It challenges our theoretical understanding of the world. But are we really certain that anything that falls into a black hole can never escape? Never say never . . .

Black Holes III:
The Mystery of the Centre

There is something paradoxical in what we know about black holes. They have now become 'normal' objects for astronomers. Astronomers observe them, count them and measure them. They behave exactly as Einstein's theory predicted a century ago, when no one dreamed that such peculiar objects could actually exist. So, they are under control. And still, they remain utterly mysterious.

On the one hand we have a beautiful theory, general relativity, confirmed in spectacular manner by astronomical observations, which account perfectly well for what the astronomers see: these monsters that swallow stars, revolve in vortices and produce immensely powerful rays and other devilry. The universe is surprising, variegated, full of things that we had never foreseen or imagined the existence of, but comprehensible. On the other hand, there is still a small question of the kind that children specialize in when adults are overly enthusiastic: 'But where does all the material that we see falling into a black hole *go*?'

And this is where things become difficult. Einstein's theory provides a precise and elegant mathematical description even of the inside of black holes: it indicates the path that material falling into a black hole must follow. The matter falls ever faster until it reaches the central point. And then . . . then the equations of Einstein lose all meaning. They no longer tell us anything. They seem to melt like snow in sunshine. The variables become infinite and nothing makes sense. Ouch.

What happens to matter that falls into the centre of the hole? We don't know.

Through our telescopes we see it falling, and we mentally follow its trajectory until it nearly reaches the centre, and then we have no knowledge of what happens next. We know what black holes consist of, both outside and inside, but a crucial detail is missing: the centre. But this is hardly an insignificant detail, because everything that falls in (and into the black holes that we observe in the sky, things continue to fall) finishes up at the centre. The sky is full of black holes into which we can see things disappear . . . but we don't know what becomes of them.

The roads taken to explore answers to this question have so far been hazardous. Perhaps, for instance, the matter emerges in another universe? Perhaps even our own universe began this way, emerging though a black hole opened in a preceding one? Perhaps at the centre of a black hole everything melts into a cloud of probability where spacetime and matter no longer mean anything? Or perhaps black holes irradiate heat because the matter that enters them is mysteriously transformed, over zillions of years, into heat.

In the research group I work with in Marseille, together with colleagues at Grenoble and at Nijmegen in the Netherlands, we are exploring a possibility that seems to us both simpler and more plausible: matter slows down and stops before it reaches the centre. When it is most extremely concentrated, a tremendous pressure develops that prevents its ultimate collapse. This is similar to the 'pressure' that prevents electrons from falling into atoms: it is a quantum phenomenon. Matter stops falling and forms a kind of extremely small and extremely dense star: a 'Planck star'. Then something happens that always happens to matter in such cases: it rebounds.

It rebounds like a ball dropped on the floor. Like the ball, it rebounds along the trajectory of the fall, in temporal reverse, and in this way the black hole transforms itself (by 'tunnel effect', as we say in the jargon) into its opposite: a white hole.

A white hole? What is a white hole? It is another solution to the equations of Einstein (like black holes are) about which my university textbook says that 'there is nothing like it in the real world' . . . It is a region of space into which nothing can enter, but from which things emerge. It is the time reversal of a black hole. A hole that explodes.

But then why do we see matter fall into black holes but do not see it immediately bouncing back out again? The answer – and this is the crucial point about what we are dealing with – lies in the relativity of time. Time does not pass at the same speed everywhere. All physical phenomena are slower at sea level than in the mountains. Time slows down if I am lower down, where gravity is at its most intense. Inside black holes the force of gravity is extremely strong, and as a result there is a fierce slowing of time. The rebounding of falling matter happens rapidly if seen by someone nearby, if we can imagine someone venturing into a black hole to see what it's like on the inside. But seen from outside, everything appears to be slowed down. Enormously slowed down. We see things disappear and vanish from view for an extremely long time. Seen from outside, everything looks frozen for millions of years – exactly how we perceive the black holes we can see in the sky.

But an extremely long time is not an infinite time, and, if we waited for long enough, we would see the matter come out. A black hole is ultimately perhaps no more than a star that collapses and then rebounds – in extreme slow motion when seen from outside.

This is not possible in Einstein's theory, but then Einstein's theory does not take quantum effects into account. Quantum mechanics permits matter to escape from its dark trap.

After how long? After a very short time for the matter that has fallen into the black hole, but after an extremely long one for those of us observing it from outside.

So here is the whole story: when a star such as the sun, or a little bigger, stops burning because it has consumed all its hydrogen, the heat no longer generates enough pressure to counterbalance its weight. The star collapses in on itself, and if it is sufficiently heavy it produces a black hole and falls into it. A star of the dimensions of the sun, that is to say thousands of times bigger than Earth, would generate a black hole with a diameter of one and a half kilometres. Imagine it: the whole of the sun contained within the volume of a foothill. These are the black holes that we can observe in the sky. The matter of the star continues on its course inside, going ever deeper until it reaches the monstrous level of compression that causes it to rebound. The entire mass of the star is concentrated into the space of a molecule. Here the repulsive quantum force kicks in, and the star immediately rebounds and begins to explode. For the star, only a few hundredths of a second have elapsed. But the dilation of time caused by the enormous gravitational field is so extremely strong that when the matter begins to re-emerge, in the rest of the universe, tens of billions of years have passed.

Is this really the case? I don't know for sure. I think it might well be. The alternatives seem less plausible to me. But I could be wrong. Trying to figure it out, still, is such a joy.

Kip and Gravitational Waves

This story begins in 1915, during the First World War, when Albert Einstein, already recognized as one of the major physicists of his time, publishes the equations of his strange theory asserting that the space in which we are immersed can be deformed like hard rubber.

He also adds immediately that it could vibrate like the string of a violin or an iron rod, and in doing so transmit *waves*. Shortly after its publication, however, he changes his mind and writes an article rejecting the existence of such waves. Then he changes his mind again and writes a further article to say that, yes, they should exist after all.

For decades, physicists are confused and debate the reality or non-existence of gravitational waves. Richard Feynman subscribes to the idea that they are real. Others disagree: if space vibrates, we vibrate with it and yet do not notice it . . .

The matter is clarified only in the sixties, forty years after Einstein's doubts: an Austro-English theorist called Hermann Bondi demonstrates that, theoretically, it is possible to boil a small saucepan of water with gravitational waves, and finally everyone is convinced. The theory predicts that space can transport vibrations similar to electromagnetic ones: ripples in space like those made by the wind on the surface of a lake.

Once this has been clarified, the question arises as to whether these waves can be observed in reality, actually running throughout interstellar space. An American physicist, Joe

Weber, constructs an enormous metal cylinder with the idea that waves from space might cause it to vibrate – and he becomes convinced that he has witnessed precisely such vibrations. But he does not manage to convince anyone else, and ends up increasingly isolated and irascible. By now, though, research into the question is fully under way.

Italy is at the forefront of this research. Eduardo Amaldi, the father of the great Roman school of Physics, senses the importance and the viability of the enterprise and promotes the Italian line of research to detect these elusive waves. Prototypes of antennae are built in Italy – first in Frascati, then later in Legnaro near Padua. They pursue Weber's idea, using large metal rods, but they also explore other ideas.

I remember as a young man, a young, aspiring student of physics, being shown by Massimo Cerdonio and Stefano Vitale in the physics department in Trento an oscillating tin with a superconducting ring inside it: the prototype of another idea for an antenna. Massimo Cerdonio went on to build the antenna at Legnaro; Stefano Vitale now leads the path towards the most spectacular international project of gravitational antennae envisaged for the future: LISA, a composite antenna made up of satellites in solar orbit . . .

But it finally became clear that the most promising technology for detecting the waves are the 'interferometers': two lasers at ninety degrees to each other which compare the lengths of two perpendicular arms. If a wave passes, one arm lengthens and the other retracts, allowing the wave to be seen.

One after another, in different countries, projects are launched to construct prototypes of similar antennae, but the sensitivity required from them is spectacularly high, way beyond the capacity of our available technology. We need to measure variations in length much smaller than an atom, over distances of kilometres. In the early nineties I was a young professor in America when Richard Isaacson came to Pittsburgh, where I was working. Richard was at the time responsible for gravitational physics at the National Science Foundation, the American agency that assigns funds for scientific research. He was in the process of making up his mind whether to invest funds in gravitational waves. The project that was being proposed aimed to detect waves within five to ten years. The two of us had dinner in a small Indian restaurant. He asked my opinion. I said that the science was sound and the project was fascinating, but like many others I was perplexed – the waves are weak, and before the technology will be capable of detecting them a considerable amount of time might have to pass. I asked him why he was convinced that it was possible to get there in a reasonable amount of time. His answer was cut and dried: faith in Kip Thorne. Kip is one of the best relativists in the world. He works at Caltech, is a world expert on black holes, neutron stars and other wonders of the universe where catastrophes of such extreme violence occur as to rock space and permit ripples of the events to reach us.

A few years later I meet Kip at a conference in India. We're sitting next to each other on a bus that is returning us to the hotel after the conference dinner. I ask him what gives him the confidence to convince Isaacson that we could detect gravitational waves. Kip pauses for a while before answering, looking me straight in the eyes. The Indian night is rushing past us. He says, 'Don't you think that we should at least try?' And I realize how high the stakes of a great poker hand in science are.

Kip now has a Nobel Prize. I reminded him of this conversation last year, after the detection had been made. His response was instant: it wasn't thanks to him, he just put his faith in Rainer Weiss and Barry Barish, spectacularly gifted experimenters.

Twenty-five years have passed since my dinner with Isaacson; twenty since my conversation with Kip on an Indian bus. The poker game was an arduous one. It was the lifework of dozens and dozens of colleagues. We won.

Thank You, Stephen

$$T = \frac{\hbar c^3}{8\pi G k M}$$

Stephen Hawking is no longer with us. We're bereft of his sly smile and his youthful irreverence, which he kept even when afflicted by old age and illness. And what an illness . . .

Only three months have passed since we lost him, but we can already try to ask ourselves, calmly and beyond our immediate reactions, what his legacy really consists of – in physics and beyond. I'm going to attempt an answer, tiptoeing as it were, out of friendship and the great admiration that I have for him.

Stephen was first and foremost an excellent physicist, one of the very best of his generation; not the greatest scientist of the century, the new Einstein or the new Newton, as he was sometimes called, with an exaggeration that he himself did not for a moment hesitate to impishly, playfully, nourish. So, I will begin with his most important scientific results.

His major discovery, the one that will be linked to his name for ever, was the demonstration that black holes behave as if they were hot: they irradiate heat like a stove. He arrived at this conclusion in 1974, with a complex and delicate calculation that skilfully combined techniques of general relativity and the theory of elementary particles. The temperature that he calculated is today known as 'Hawking temperature', and it depends on the size of the black hole. The larger the black hole, the colder it is. Hot black holes are therefore the smallest. This result provoked a big surprise in the seventies, bringing Stephen, who was barely thirty years old, renown among theoretical physicists. Up until then, no one had expected that a black hole could have a temperature. Not even Stephen himself, until he had completed the calculation.

The heat that irradiates from black holes is today called 'Hawking radiation'. It has never been observed, and it will be difficult to observe it any time soon because it is so weak. But its existence has been vouchsafed in many different ways, and it is accepted as plausible by the vast majority of scientists.

Why is 'Hawking radiation' important? Because it is a phenomenon that involves both the structure of spacetime and quantum mechanics. This makes it an important indicator with regard to one of the great open questions in contemporary physics: the search for a theory of 'quantum gravity', which is to say a theory that describes all the 'quantum' aspects of space and time. Hence much contemporary research uses Hawking's result or seeks to develop it. The research group in which I work, for example, is currently trying to use a possible theory of quantum gravity to calculate what happens to a black hole *after* being consumed by irradiating Hawking radiation.

There is a beautiful formula that summarizes Hawking's result. It is the formula that gives the temperature, as a function of the mass M of its surface. This is the extremely simple formula given above: $T = \hbar c^3 / 8\pi G k M$.

The beauty of this formula lies in its simplicity, but above all in the fact that it combines the four fundamental pillars of physics: the Boltzmann constant k, which is the root of thermodynamics, the speed of light c that characterizes relativity, the Newton constant *G* that characterizes gravity, which is to say the structure of spacetime, and the Planck constant ℏ on which quantum mechanics is based. There is no other formula that puts together so elegantly all the fundamental pillars of our physics. No wonder Stephen asked for this formula to be carved on his headstone.

Among Stephen's minor results, there are two that stand out as particularly relevant. As a young man, in collaboration with the great English mathematician Roger Penrose, he demonstrated that Einstein's theory predicts that the universe emerged from a 'Big Bang': a singularity where the theory no longer works. This conclusion had been reached previously only by assuming, not very realistically, that the universe is completely homogeneous. Penrose and Hawking's theorem showed that this simplification was not necessary – a result that helped to make the Big Bang much more plausible.

Stephen returned to the question of the Big Bang in the eighties, seeking to show how a quantum theory can effectively describe the birth of the universe. He constructed a fascinating intuitive model of quantum gravity and applied it to the inception of the universe. The model still inspires current research in quantum gravity.

There is a thread that connects these results. As a young man, Stephen had become passionately devoted to Einstein's great theory. At the time, the applications of it were few, and the research mathematical. The most spectacular predictions of physics, such as black holes and the Big Bang, were still thought to be esoteric and unsound. Penrose had bolstered them by showing that black holes undoubtedly form when sufficient amounts of matter become concentrated, and Stephen had the idea of using Penrose's technique for studying the origin of the universe. The idea was that the birth of the universe was something like the collapse of a black hole seen backwards in time.

Having clarified that inside black holes and in the primordial universe Einstein's general relativity became insufficient, Stephen was prompted to start to consider quantum effects. In this way he arrived at Hawking radiation. Then, in subsequent years, he attempted to use quantum mechanics fully in order to rethink the beginning of the universe in terms of quanta. All these problems are still unresolved. But in contemporary discussions of them it is not uncommon to hear reference to the name Hawking, or to one of his ideas.

My summary hardly exhausts Hawking's theoretical activity, but I hope to have given a basic sense of what he contributed to physics.

I believe, however, that his greatness lies elsewhere.

His real greatness was his humanity, his character. Tied to a wheelchair, he progressively lost control of all the muscles in his body. The last time I saw him, in Stockholm, he could barely even move his eyes. He communicated by moving them: an electronic system read his eye movements with a small camera, and thanks to this Stephen was able to control a computer to put letters laboriously in line, to construct words that were then spoken by a vocal synthesizer. It was painful to watch him undergo this exhausting and extremely slow process.

And yet the voice of that synthesizer reached the entire world. That so distinctive metallic voice which Stephen somehow managed to make his own, turning it into the almost natural medium of his brilliant intelligence and of his irony. He never lost heart. He continued to produce physics of quality, even as the condition of his body continued to deteriorate. In seemingly impossible circumstances he managed to write a book that became phenomenally successful. In the thirty years since it was published it has sold more than 10 million copies, and it continues to be read. With this book he spoke to young people throughout the world, amazing and inspiring them to study the universe.

For all the extreme misfortune of his condition and disability, Stephen was also the beneficiary of not a little luck: blessed with exceptional intelligence, he was born into an excellent family of English intellectuals and received an education of the first order. The progress of his illness was also much slower than was initially and drastically predicted. His value as a scientist, and then his fame, allowed him to achieve things that others with similar conditions could only dream of. But even when taking all this into account, Stephen, with his rather insolent air of being an untouchably youthful spirit, has given the world an extraordinary lesson in humanity. A lesson in love for life, in intelligence, and in unquenchable curiosity.

The day after our meeting in Stockholm, where communication with him was so difficult and heartbreaking, Stephen gave a lecture in a huge theatre in the city. He was everywhere surrounded, to an incredible degree, by young people hanging on his every word. He arrived onstage with his legendary smile, his legendary wheelchair, and set the prerecorded lecture running with a movement of his eyes. He talked in it about ultimate attempts to understand the future of black holes, made jokes, gently mocked the French, played with the meaning of life, irreverently, rebelliously, a smile on his lips accompanying every phrase. The enormous audience was spellbound. His last words were still an undaunted declaration of love for life, but as always with a play of ambiguity: you can indeed escape from black holes.

Stephen was certain that life does not continue in any other form after death. Like many scientists, he was fond of using 'God' for emphasis and effect, but was a confirmed atheist, without ambiguity or uncertainty, and he said so clearly and without hesitation. It wasn't any kind of transcendence that he found consolation in or drew his strength from. He was imprisoned by the most debilitating of illnesses, linked to the rest of us by an ever-thinner thread. Yet he continued to live until the end with a burning intensity – to joke, to speak to the whole world, to communicate happiness and joy, moving new generations to follow him with his enthusiasm. Is this not an extraordinary lesson in life for all of us whingers? Is this not the infinitely precious gift that Stephen has left us? The irresistible luminous force of life, of curiosity, of thought, of intelligence.

Now that thinnest of threads has been cut. Before disappearing for ever, as happens to all things, dissolving into the immensity of the boundless cosmos that he loved, Stephen still remains for a while, alive and active in our science, in our memory, in our affections, in our thoughts. Thank you, Stephen.

Roger Penrose

A few days ago I had the pleasure of meeting Roger Penrose, the great mathematician from Oxford, who was passing through Italy for the Festival of Science in Genoa. Penrose is a polyhedral intellectual. Readers know him for several books, among them the dense and wonderful *The Road to Reality*, a great panorama of contemporary physics and mathematics, a popular work that is not easy and that shines with intelligence and profundity on every page.

Among the main contributions made by Penrose to our knowledge of the universe are theorems showing that Einstein's theory implies that the universe we see originated from a Big Bang. In the field of pure mathematics, he is better known for his study of 'quasi-periodic' structures, tessellations composed of a few elements that can be repeated to infinity but that, however, are not periodic: they never repeat identically. They are also known as 'quasi-crystals' and exist in nature, but they have also been used in fields that range from the design of floor tiles to a children's game devised by Penrose himself. And currently it is possible to admire Penrose's formulas even in a contemporary art exhibition: Penrose has developed a method of calculation based on long sequences of drawings, and Luca Pozzi – a brilliant Italian artist very receptive to science, has exhibited his formulas at Grenoble in a small exhibition dedicated to art and science.

Today Penrose is an eighty-year-old gentleman who has retained a youthful air and eyes still enchanted by the world.

He jokes about how one's memory goes, and about the house keys that he locked inside when he closed the door that morning, leaving early for Italy. But his mind is as lucid as anything, and when he talks about his latest idea, presented for a general reader in his most recent book, *From the Big Bang to Eternity*, he becomes enthused.

His idea is that by looking at the sky, perhaps it is possible to see, or rather to have perhaps already seen, traces of events that occurred before the Big Bang. These traces could be immense concentric rings in the sky that may be glimpsed in the 'cosmic background radiation', the weak residual radiation of the Big Bang which fills the universe. Think of the waves that persist in a pond after a stone has been dropped into the water, forming concentric circles of increasing size. In this case the pond is the entire universe, and the stone that fell into the water a collision between colossal black holes that occurred *before* the Big Bang . . .

We have recently discovered that the universe is expanding at ever-increasing speed. What will happen in the distant future? The clusters of galaxies will recede from each other at increasing speeds, stars will be extinguished, everything will be reduced to a few black holes and light waves wandering in an ever more boundless and glacial space. With the lapse of eons of time, black holes themselves will end up evaporating and there will be nothing left but a universe of waves of light racing through nothingness for all eternity. 'A desolate and terribly boring prospect,' Penrose jokes, adding, 'But fortunately the waves of light won't get bored.' That sounds like a joke too, but it is in fact an acute observation. As Einstein was the first to realize, in fact, the faster we move, the slower time passes for us. If we could make a journey at extremely high speed, we would find on our return

$$Q^{abc}_{fg} \rightsquigarrow \quad , \qquad Q^{abc}_{fg} - 2Q^{bca}_{gf} \rightsquigarrow \quad - 2$$

$$\xi^a \lambda^{(d}_{ab[c} D^{e)b}_{fg]} \rightsquigarrow \tfrac{1}{12} \qquad \lambda^a_{bcd} \rightsquigarrow$$

$$D^{ab}_{cd} \rightsquigarrow$$

$$\delta^a_b \rightsquigarrow \qquad \xi^a \rightsquigarrow$$

$$= | \, | - X, \qquad = | \, | \, | + X\!\!\!/ + X\!\!\!\backslash - X| - |X - X\!\!\!\!X$$

$$= | \, | + X, \qquad = | \, | \, | + X\!\!\!/ + X\!\!\!\backslash + X| + |X + X\!\!\!\!X$$

$$= | \, | - X, \qquad = | \, | \, | + X\!\!\!/ + X\!\!\!\backslash - X| - |X - X\!\!\!\!X$$

$$\xi^{[a}\eta^{b]} \rightsquigarrow \tfrac{1}{2} \quad , \quad \xi^{[a}\eta^{b}\zeta^{c]} \rightsquigarrow \tfrac{1}{6} \quad , \quad \xi^a \rightsquigarrow \quad \eta^a \rightsquigarrow \quad , \quad \xi^a \rightsquigarrow$$

Penrose calculations

that our contemporaries were much older than us. The closer we get to the speed of light, the greater the effect would be. If we could travel at the very speed of light, time for us would stop. It would stop flowing altogether. But light obviously moves at, well, the speed of light – so for light time never passes at all. In this sense, light 'won't get bored'.

A universe in which nothing existed but light would be a universe in which nothing could 'perceive' the passage of time. Time would literally no longer exist. Not only would this be the case, but if only light existed, we could not even measure spatial distances. The universe of our most remote future, Penrose observes, would be a universe that we could describe as immensely large and immensely durable, but in reality it would be a universe devoid of duration and without dimensions.

But just at the start of the Big Bang, an instant before beginning to expand, the universe found itself in a situation of this kind: without duration and without dimension. And it is here that Penrose makes his amazing suggestion: what if the most remote future of the universe was in fact the initial Big Bang of a new cycle of the universe? In both cases there is neither duration nor distance: an immensely expanded universe is in reality the same as an infinitely small universe. We can imagine a 'recycling' of the universe in which the scale of distances vanishes and is redefined. Perhaps the immensity of the future universe is none other than the microcosm of the universe at its birth, just seen 'at a different scale', and our very own Big Bang is nothing but the infinite future of a preceding universe.

Can we seek to confirm these hypotheses? Well, Penrose observes, the very last events before the disintegration of time would have been the last great collisions of black holes before their final evaporation. Is it possible that these collisions left a trace? That trace might consist of some slight ripple in the sea of final light. Some great circle expanding through the cosmos, centred on the last great events of the universe. These great circles may have gone through the phase in which the universe recycled itself, restarting from a new Big Bang. If our universe was indeed the product of such evolution, we should be able today to see in the heavens these great circles produced before the Big Bang. This is Penrose's audacious hypothesis.

It has an extremely speculative flavour to it. But last year the astrophysicist Vahe Gurzadyan of the Physics Institute of Erevan, in Armenia, announced that he had found circles of this type in the sky, by analysing the data accumulated over the years on cosmic background radiation, collected by the observatories WMAP and BOOMERANG. The observation is not very clear, and its interpretation is very controversial. There are objections that we could be dealing with random fluctuations: after all, it is easy to 'see' shapes in the clouds. The question remains open.

I don't know how it will end. Perhaps the circles will prove to be illusory. But I don't think the idea of looking for clues relating to pre-Big Bang events will go away. In any case, there are two important lessons that can be extrapolated from this idea. The first is the precise grounding in observation to which Penrose remains proudly faithful. It does not matter how out there the idea seems; what is important is that it is anchored in the possibility of being verified: we look for the circles in the sky. This is good science. And it is a healthy counterbalance to the many research programmes that continue for decades without producing any precise predictions, stuck in the infinity of that sad limbo where theories can neither be verified nor proved to be false.

The second lesson, only apparently at odds with the first, is underlined by Penrose himself, with a disarming smile: 'I can't bear the thought that the universe is sinking towards an infinite future of frozen death.' This expresses a feeling, a vague intuition, an emotional need. But science, even the best science, can also originate in this way – from the heartfelt rejection of a future that is too boring to countenance. If they are anchored in the concrete possibility of being verified or not, the best ideas can be, and indeed have often been, the fruit of wholly irrational intuitions, almost of a vague empathy with the nature of things. Roger Penrose, at eighty, remains a true master of science.

Dear Baby Jesus

Dear Baby Jesus, this is my little Christmas letter for 2015.

I feel, dear Jesus, that I am in credit with you. Because you and I had a problem once, some time ago. Remember?

I was an infant myself at the time. Just like you. I had a mother and a father who loved me, just like you, and they were very honest with me, as I suppose were yours. They did not encourage me to believe in nonsense. They did not tell me that children are delivered by a stork, or that everyone in the world is good, or any other outright fibs of this kind. They told me the truth. Moreover, they told me that one should never tell lies. I was very proud of these parents who seemed always to tell the truth.

And yet they did tell me one lie. At Christmas I would find presents under the Christmas tree, and Mum and Dad would tell me that those presents had arrived magically from the sky, and that they had been brought by you. They would tell me that Baby Jesus had magically delivered the Christmas presents. I really believed this, I would be full of excitement waiting for Christmas, waiting for the magic to occur. Then one day I was told that it was really them who had bought the presents for me and placed them under the tree. I burst into tears.

I felt that I had been toyed with by my parents – and perhaps even more so, Baby Jesus, by you. So you weren't after all able to deliver presents on your own? You needed help?

Why did they have to tell me such a stupid lie as this? The

truth, as I eventually came to realize, was anyway much better than the fib: the presents had been bought by them, my parents. Because they loved me and wanted to make me happy. Isn't it a beautiful thing, that my mother and father wanted to see me happy? Isn't it wonderful that they loved me so much? This was the real magic, the real emotion. Why hide it behind the stupid ruse of Baby Jesus, or Santa Claus, or who knows what other gift-delivering angel or supernatural entity?

In short, all the presents that you were meant to have brought to me were never delivered. That's why I feel in credit with you.

So now I'd like to redeem this debt, with a little interest into the bargain. What follows, dear Jesus, is what I'm asking from you this year.

Instead of bringing so many presents to the silly children of well-off families (children such as I was), something that in reality you could not be accused of doing anyway because it is done by the parents, please go to Syria instead and stop the bombs from raining down on innocent people. Go to India and bring food to those who do not have enough to eat. Go to Europe and change the selfish minds of all those who jealously guard their wealth and do not want foreigners around, and go to Africa and give something to the millions of people who live in utter misery. Go around the world – why not? – and stop this wave of fear, hatred and distrust that has affected everyone in 2015. Christians who frown on Muslims, Turks who fire on Russians, Russians who fire on Syrians, Saudis bombing Yemenites, Europeans hitting out on all sides. People being killed everywhere; everybody eager to go to war.

Baby Jesus, if you can, please stop this escalating madness.

164

Otherwise, if you can't do any of this, please stop deceiving us with saccharine illusions, and leave to men and women of good will the responsibility of bringing real, useful gifts to the world. Just as my parents used to do with me.

Leave us in peace. Because it is only men and women of good will who are able to alleviate suffering in this world. You, dear Jesus, are as incapable of doing this as you were of bringing me a present.

Certainty and
Global Warming

Even as world leaders are meeting in Paris to seek a difficult accord on limiting the damage being done by global warming, voices are being raised all over the world to minimize the problem. They argue that nothing is certain about the climate. I believe that those taking this line cannot possibly realize the extent of the damage that is being done to all of us. I think they need to be opposed simply and clearly.

Whoever says that we cannot have absolute certainty about the future climate of the planet is stating an obvious truth. But it is obtuse to say that a danger is not grave because it is not mathematically certain that it will happen.

If we discover a bomb that has remained buried beneath what is now a children's playground, we do not leave it there because 'it might not explode'. If a fire breaks out in a cellar, a reasonable person looks for a fire extinguisher, calls 999, escapes from the building. Whoever says: 'But there is no certainty that the fire will spread, therefore let's calmly carry on with breakfast,' is a cretin. And yet this is precisely the attitude taken by those who argue that the problem is not serious, because we have no certainty regarding the climate.

At the risk of stating the obvious, I will attempt to summarize the situation.

It is a fact that the Earth is currently subject to an unusually accelerated degree of warming. It has become clear (as it was not just a few years ago) that global warming is being significantly increased by human activity, especially by the emission of carbon dioxide. Predicting the future of the climate is difficult. The projections indicate that if no intervention is implemented, the increase in the planet's temperature will reach four to five degrees this century. This would lead to catastrophes in the coming decades. In the past, changes of temperature of this magnitude have caused extinction events.

For the Earth, these are small fluctuations comparable to many others; for humanity, it is set to be disastrous: it could mean the flooding of cities near the coast and on large plains, vast desertification, the collapse of agricultural production, famine, widespread hunger, hurricanes, conflict breaking out everywhere.

We are not talking about the welfare of polar bears. We are talking about the future of our children.

The emission of carbon monoxide due to human activity continues to exacerbate the problem. Coordinated action by humanity to reduce emissions could succeed in reducing warming by two degrees Celsius, thus limiting the most damaging effects, if not all of them. This is the analysis of the IPCC, the United Nations Intergovernmental Group for Climate Change. And there is universal consensus on this from every serious institution on the planet. Those who disagree are no different from those who try to argue that dinosaurs never existed, or that the Earth is flat.

This is the situation that we are in. We have no certainties. We might be wrong. But we have to take a decision. We can decide to ignore the alarm bells and continue regardless, on the basis of the fact that 'we are not altogether sure'. This is the attitude of those who would leave the bomb beneath the playground because it might not explode. The whole world has become convinced that there is a serious risk, and those voices that generate confusion by denying the fact simply make things even more complicated for those who, with difficulty, because the problems are so complex, are trying to mitigate the danger on behalf of us all.

Churchill and Science

Churchill was the first British prime minister to appoint a scientific adviser, as early as the forties. He had regular meetings with scientists such as Bernard Lovell, the father of radioastronomy, and loved talking with them. He promoted with public funds research, telescopes and the laboratories where some of the most significant developments of the postwar period first came to light, from molecular genetics to crystallography using X-rays. During the war itself, the decisive British support for research, promoted by him, led to the development of radar and cryptography, and played a crucial role in the success of military operations.

Churchill himself had a scientific grounding that was hardly extensive but was nevertheless sound. As a young man he had read Darwin's *On the Origin of Species* and studied an introduction to physics: the essential things, we might say. He followed scientific advances with considerable interest, to the extent that in the twenties and thirties he wrote articles on popular science. For good or ill, he was the person who signed the fateful note sent to the father of quantum mechanics, Niels Bohr, in Copenhagen, inviting him to flee Nazi-occupied Denmark to join the Allies, and to initiate the atomic programme.

In *Nature* magazine, the American astrophysicist and author Mario Livio describes an unpublished text from 1939, revised in the fifties, in which Churchill discusses a scientific issue of great relevance today: the possibility that life may exist elsewhere in the universe, on planets similar to Earth. Churchill's analysis is surprisingly lucid, demonstrating an uncommon capacity to use scientific language. Anticipating the conclusions that would be reached within the scientific community in coming decades, Churchill identified the elements that would enable forms of life similar to those on Earth to develop elsewhere: planets with the required distance from their mother star, allowing them to maintain temperatures within that small band width in which water is liquid, and with a large enough mass to sustain a sufficiently dense atmosphere.

Then there is one particularly impressive passage: Churchill notes that the most credible contemporary theory on the formation of planetary systems – the close encounter of two stars – makes the conditions required highly improbable, and hence life rare; but he notes further that this conclusion depends on the validity of the theory regarding close encounters, and no one can say for sure whether the theory is correct. Not only was this great statesman capable of valuing the importance of the scientific knowledge that he encountered, but he also had an acute sense of its margin of uncertainty. The theory of close encounter or close passage did in fact turn out to be mistaken, and today we know that planets form in a different way, from the aggregation of much smaller fragments. Churchill's main conclusion is close to the one that we hold today:

With thousands of millions of nebulae (galaxies), each one containing hundreds of millions of suns, the probability that there is an immense number containing planets where life is possible is high.

The comment that follows captures perfectly, for me, something of the English spirit:

> As for me, I am not so terribly impressed by the successes of our civilization as to believe that in this immense universe we represent the only corner where there are living and thinking beings, or that we may be the highest level of mental or physical development that has been reached in this vast expanse of space and time.

Churchill clearly saw the limits of science. 'We need scientists in the world,' he writes in 1958, 'but not a world for scientists.' And he adds: 'If, with all the resources that science has put at our disposal, we are still unable to defeat hunger in the world, we are all culpable.' But he was profoundly aware of the central role of scientific thought for humanity; of the importance of politically supporting it, of listening to it, and of using it. Above all, he was aware of the great advantage it offered in allowing him to come to political decisions based on the facts – the simple secret that has contributed significantly to British and then American political supremacy in the last two centuries. Churchill knew how to think with the clarity of scientific intelligence.

Traditional Medicine and UNESCO

Several remedies used by modern medicine derive from traditional ones: modern science has found a way of evaluating them, and has recognized the efficacy of various traditional medical practices. However, a great many other traditional medicines and medical practices have been subject to the same evaluation and have been found to be either inefficient or actually detrimental to health. One example among countless others is the bloodletting extensively used by traditional European medicine for centuries, before it was realized that it was extremely harmful.

It is the sacrosanct right of anyone who is ill to choose their cure, or even to opt for no cure if that is their wish. However, there is an equally sacrosanct obligation upon anyone, including the press but above all upon institutions, if they do not wish to lose public confidence and their reason for existing in the first place, not to compromise the reliability of genuine medical advice given to citizens. If Mr Y wishes to resort to bloodletting to cure himself of something, that is his business. But if an institution recommends, guarantees or promotes in any way the use of bloodletting, in the name of a 'respect for traditions', this institution is failing in its duty. It is also committing a criminal act.

This is not an academic question. A dear friend of mine, Simonetta, died before she was thirty years old, leaving behind two motherless children, because instead of relying on modern medicine to cure a breast tumour – a pathology that today has extremely high survival rates – she trusted instead in 'traditional medicine'. Whoever has given credit to the supposed efficacy of traditional medicine, and therefore influenced my friend and many others, should have her death and her children's lack of a mother on their conscience – together with thousands of other, similar deaths.

When we also discover that there is a highly lucrative trade in alternative medicine worth millions, profiting from the hopes of the afflicted and refusing any independent evaluation of the efficacy of their products and practices, then we know that we are dealing with a serious problem.

Recently, the international press has been reporting an ongoing argument between India and China, who are litigating the paternity of traditional Tibetan medical practices. They have each demanded that UNESCO should register these practices exclusively in their name in its list of the Intangible Cultural Heritage of Humanity.

UNESCO's list of cultural patrimony cannot be used to recognize everything that is 'traditional'. If this was the case, it would have to include slavery, forced child labour, the right of husbands to beat their wives, and human sacrifice.

The list recognizes that part of cultural heritage which we all wish to preserve and to see thrive. Traditional medicine, even if it has given us a useful legacy, and one yet to be fully and properly explored, does not belong to a heritage that we want to preserve per se, because it is also full of remedies that an independent scientific evaluation would find to be inefficient or harmful.

As much as the actively harmful ingredients and practices of traditional medicine, its inefficiency can be lethal. It was thanks to the uselessness of a traditional South American cure that my friend died from her breast tumour.

The letter 'S' in UNESCO stands for 'Scientific' (United Nations Educational, Scientific and Cultural Organization). A UNESCO recognition of traditional medicine through its inclusion in the list of the Cultural Patrimony of Humanity would be something that India, or China, could boast about. It would constitute a blanket recognition of the value of ancient practices by the foremost international authority for education, science and culture. It would provide a fabulous business opportunity for many. UNESCO would lose all credibility. It would be nothing short of an act of criminal irresponsibility.*

* In 2018, Tibetan medicine *was* added by UNESCO to the list of the Intangible Cultural Heritage of Humanity.

The Infinite
Divisibility of Space

A good philosopher who concerns himself with science brings new light to the subject and adds conceptual and historical depth to the big scientific questions. This was my first thought after reading *The Paradoxes of Zeno*, a sophisticated short book by Vincenzo Fano, an Italian philosopher who teaches at Urbino and who is equipped to deal with science.

Zeno was also an Italian philosopher who was interested in science. He taught in Elea, in what is now Cilento, in the province of Salerno, 2,400 years ago. Fano's book places his work in a contemporary light, close to questions of theoretical physics that are at the heart of my research.

Zeno has passed into history on account of several arguments known today as 'Zeno's Paradoxes' – arguments which, in the words of Bertrand Russell, are 'immeasurably subtle and profound'. They came at the beginning of the great era of Greek thought. In the preceding decades, the philosophers of the naturalistic school of Miletus, in what is now Turkey, had begun to appreciate that things are not necessarily as they seem, but that their nature can be investigated and understood by reason. In Italy, the philosophers of the school of Elea, following in the footsteps of Parmenides, had taken this intuition to a radical extreme, maintaining that true reality is *only* what can be reconstructed by reason, therefore denying reality to mere appearances. Zeno presents his 'paradoxes' as arguments to show that movement is inconceivable: hence movement is just one of the false 'appearances', and not a true reality. A daring idea.

The most famous of these paradoxes is that of Achilles and the tortoise. Achilles and the tortoise challenge each other to a race, with Achilles reckoning that he runs ten times faster than the tortoise and conceding to his opponent a hundred-metre advantage. How long will it take for Achilles to catch the tortoise?

Before reaching the tortoise, Achilles will have to cover the one hundred metres he has given as a head start, and this will take a certain amount of time. In the meantime, the tortoise will have covered a certain amount of ground: ten metres, say. Achilles will have reached the tortoise's point of departure, but the tortoise will be ten metres ahead. Before reaching the tortoise, Achilles will also have to cover these ten metres, which will take more time. But, meanwhile, the tortoise will have covered a bit more ground, and so on to infinity. Hence before reaching the tortoise, Achilles will have to cover *an infinite number of portions of space*, and each one will take time. Hence Achilles will require *an infinite number of times* – that is, according to Zeno, an infinite amount of time.

In other words, he will never catch up with the tortoise. If we do actually see Achilles catching up with and overtaking the tortoise, after we have demonstrated that it is not possible, it follows that what we are seeing is an illusion. This is Zeno's argument.

But the reasoning is wrong. The error is to think that a sum of infinite intervals of time must result in an infinite time. This is not true. To see this, we need only think of taking a piece of thread one metre long and cutting it at fifty centimetres. Then cutting it at twenty-five centimetres, then at twelve and a half, and so on, always adding to the cut pieces a length half of the one before it. If we continue this exercise to infinity, will the cut pieces add up to a thread of infinite length? Obviously not, because taken all together the segments cannot possibly be longer than the metre that we had at the outset. Hence the sum of infinite lengths can easily be a finite length, when the lengths added together are gradually shorter. In the same way the sum of infinite times can very well become a finite time. Zeno's reasoning is mistaken at the point when he deduces that the sum of infinite times covered by Achilles must add up to an infinite time.

This consideration seems to settle the matter. It is a solution that twenty-four centuries ago perhaps escaped the best minds, because familiarity with infinite sums only developed later. They began to be understood a couple of centuries later, by Archimedes in Italy, but clarity was achieved regarding them only in the modern era: mathematicians today call them 'convergent series'. Complete clarification was achieved gradually, but there can be no doubt now that Zeno's argument about Achilles and the tortoise is misconceived. That's why, before reading Fano's book, I had always considered the paradoxes of Zeno to be of little interest.

But then why did Bertrand Russell, who certainly knew something about mathematics and was far from naïve, consider those paradoxes to be 'immeasurably subtle and profound'? And why has Vincenzo Fano, an acute philosopher, dedicated a book to Zeno at this time? The point is this: are we really certain that the solution I have just given is the right 'physical' response to the question posed by Zeno? Is this really what happens when Achilles races after the tortoise? Does he really cover an infinite number of segments of ever-decreasing length?

Let's put the question another way, still following in the

steps of Zeno. At school we learned that space is a set of points. But a point has no extension. Not even two points have extension, or three. As a matter of fact, however many points I add together, I will always have something without extension. So how can I obtain an extended space by piling points together? Fano's book, which is didactic and intelligent, erudite and exhaustive, lays out before us the many difficulties entailed by the concept of continuous space, the theoretical acrobatics required to get round them, the reflections and the objections that these difficulties have generated over the centuries. And he brings us to the threshold of today's version of the problem.

In order to understand the profundity of that problem, let's follow another historical thread. Zeno had a friend called Leucippus. Leucippus was the first philosopher to propose an idea that was to have a brilliant future: the atomic hypothesis. Motivated by the difficulties of the idea of divisibility ad infinitum, on which his friend Zeno was so insistent, Leucippus proposes the idea that matter is made up of indivisible units. The tremendously important development of this idea will fall to his disciple Democritus, one of the greatest philosophers of all time.

Those who study Democritus consider the loss of his texts, probably censored during the 'devout centuries', to be one of the greatest tragedies to have befallen world culture. Perhaps the world would have been a better place if, instead of getting rid of the works of Democritus and preserving those of Aristotle, our forefathers had lost all copies of Aristotle's books and managed to preserve those of Democritus instead.

Well, Leucippus and Democritus explore the idea that matter may *not* be divisible ad infinitum. It is not possible to divide a drop of water into an arbitrarily large number of increasingly smaller drops. There is a minimal unit of water: a water molecule. According to Democritus, the universe in its variety and complexity may be understood by being thought of as made up of indivisible units of matter, which he called 'atoms', and by their dance in space. The vision offered by Democritus, that we know above all in the version in verse given to us by Lucretius, inspired the birth of modern science and has been splendidly confirmed in recent times – and it is the one that every child learns at elementary school today: matter is made up of atoms: it cannot be infinitely divided.

And what about space? The original problem posed by Zeno, on the other hand, that of the infinite divisibility of space, is still very much open. What's more, it is at the centre of theoretical research in fundamental physics. The physics of Newton assumed that space was real and infinitely divisible. During the course of the nineteenth century, mathematicians developed refined theorems to account for the peculiarities of the continuum. These provide, as I've already explained, a possible solution to Zeno's paradoxes, but it is an abstruse one. In the ancient and profound arguments of the philosopher of Elea there remains something disquieting that is properly highlighted by Fano. What remains is the physical question posed by him: is space and is time *in reality* infinitely divisible? And what if this proved not to be the case?

Is it reasonable to think of physical space preserving its continuous structure, all the way down to infinity? To infinity, plunging into an abyss of the ever smaller, is an awful long way down . . .

Today, indeed, there are indications that the correct solution to the paradoxes of Zeno is not to be found in the continuum. The insight of Leucippus and Democritus may prove to be right not just for matter but for space itself.

Twentieth-century physics has demonstrated the relevance to the structure of the universe of three physical constants: the speed of light, the constant of universal gravity and the Planck constant that establishes the scale of so-called quantum phenomena. By combining these constants we obtain a length, called the 'Planck scale'. It is a very small scale indeed (a decimillionth of a billionth of a billionth of the nucleus of an atom), but one that is finite. At this scale we expect 'quantum' phenomena, and the most typical 'quantum' phenomenon is granularity. Electromagnetic waves, for example, behave like a swarm of particles: the famous photons or 'quanta of light'. In the same way it is reasonable to expect that space will also show aspects of granularity at the Planck scale. Space might well be made up of elementary 'atoms of space', providing a limit to divisibility.

This granularity of space is a key element in various theories that are currently being developed. The theory in which it is studied most explicitly is 'loop' quantum gravity, in the context of which my own research unfolds. In loop theory, a centimetre of space is not continuous: it is a collection of a very large but finite number of 'atoms of space'. Hence Fano's essay on Zeno inserts itself not just into historical-philosophical debate, but also offers arguments and points of reflection for theoretical physics. The conceptual complexities entailed in the notion of continuous space, well described by Fano, may not be the best way to think about the real world.

If loop theory is correct, Achilles will not have to take an infinite number of steps in order to overtake the tortoise. The bounds required of our hero may be many, but their number will be finite.

Ramon Llull: *Ars magna*

In 1274, at the end of a long hermitage on Puig de Randa, a mountain on the island of Majorca, Ramon Llull conceives – by divine revelation, he claims – the idea for a great work that would become the core of his life and its main objective: the creation of a complex system that he calls his *Ars magna*: his 'great art'. Llull's 'great art' is a strange and complex system oscillating between metaphysics and logic, expressed in the form of tables, graphics, moving paper circles that may be rotated and superimposed to generate arbitrary combinations of fundamental, elementary concepts. With this system, Raymond Llull – or Lully, as he is called in the English-speaking world – intended to bring order to the world, and to convert Jews and Muslims to Christianity.

These objectives, it would be fair to say, he decidedly failed to reach. But the influence of his peculiar system has been huge. Both Giordano Bruno and Montaigne, two of the thinkers who have had most influence upon modernity, drew inspiration from Llull. But it was above all Leibniz who sought to sift the wheat of Llull's 'great art' from its chaff, to cleanse it of its medieval aspects and attempt to extract a universal rational language, dubbing it 'combinatorial art', with the objective of translating rationality into calculation.

A direct application of this idea is the first calculating machine, designed by Leibniz and acknowledged as the progenitor of all modern computers. But the same idea is also at the root of modern developments in logic conceived as a universal grammar of rationality, from Frege to logical positivism. The art of Llull is the deep root of much modern thought and technology, and has made this great Catalan intellectual one of the most original and influential voices to come out of medieval Europe. One technical tool central to the physics that I work on, to take just one small example of that influence, is the graph: an image that codifies the way in which a certain number of elements are connected with each other. Graphs were invented by Llull.

At the root of the curious power of combinatorial art there is a simple fact. The greatest Persian poet, Ferdowsi, conveys it well in a famous legend in his epic *The Book of Kings*. The ingenious inventor of the game of chess, a man called Sissa ibn Dahir, gifted the game to a great Indian king. Overwhelmed by admiration and gratitude, the king asks the wise inventor how he could reward him, and he answers in the following way: 'Give me a grain of wheat for the first square on the chessboard, two for the second, four for the third, and so on, until we have doubled the amount, using every square on the board.' The king is amazed by such a modest demand and immediately orders that it be granted. So imagine his astonishment when his attendants return, and tell him that there is not enough grain in all the granaries in the kingdom to satisfy the wise man's cunning demand!

The calculation is easily done: just for the last square on the board, the sixty-fourth, it is necessary to have a number of grains equal to two multiplied by itself sixty-four times, and this amounts to 18 billion billion grains. If a single grain weighs a gram, then this would be 10,000 billion tons of grain. And this is only for the last square! Dante, in canto XXVII of his *Paradiso*, uses precisely this legend to indicate an inordinate number: 'And they were so many that

Images from Llull's *Ars magna*

their number was more than the doubling on the chess squares.'

What significance does it have – the fact that a number so small can give birth to one so large? It means something simple: the number of combinations is generally much larger than we instinctively imagine. By combining a few simple objects you can obtain an unexpected vastness of things, and these can be arbitrarily varied and complicated. It is not just the number of combinations that is astonishing: it is also their variety. Think of the nature that surrounds us. Physics has helped us to understand that all that we see is generated by

183

scarcely more than four particles which interact through a few elementary forces. The few different pieces of this simple Lego set produce forests and mountains, star-studded skies and the eyes of girls.

But the extent of what *could* exist is even greater than the immense extent of what *does* exist. Think of the proteins that form the structure of all living things. A protein is more or less a sequence of a few dozen amino acids. There are twenty amino acids. Say we want to produce all the possible proteins, in order to study them: this will allow us to understand all the possible structures of living matter, and even to anticipate the evolution of life on Earth . . .

But there is a problem: the combinatorial calculation is easily done: the possible combinations of about twenty amino acids forming chains of a few dozen elements are so numerous that even if we managed to produce a different protein every second the entire life of the universe would not be sufficient to produce more than a small fraction of all the possible proteins . . . In other words, the space of the possible structures of life is still almost completely unexplored, not only by us, but by nature itself.

The first intuition of the immensity of space opened up by complexity had already been had by Democritus, twenty-four centuries ago. Democritus had understood that the whole of nature could be made up of only atoms; and that, as he puts it, it is the combinations of atoms that generates the complexity of nature, 'just as the combinations of the letters of the alphabet can generate comedies or tragedies, epic poems and satirical plays'.

Our intuition baulks at the immense numbers and the endless variety generated by combinations. Like the king in the Persian story, it seems impossible to us that from the combination of simple things, so many other things and such complexity can be born. It is for this reason, I believe, that it seems inconceivable that things as complex as life, or our own thought, can emerge from simple things: because we instinctively underestimate simple things. We never think they are capable of much. Numbers generated by grains of wheat and a chessboard surely cannot empty all the granaries in the kingdom. And yet this is so.

Our brains contain approximately 100 billion neurons; each one of these is tied to other neurons by conjunctions, the synapses. Each neuron has several thousand synapses. Hence each of us has in our head hundreds of thousands of billions of synapses. But it is not this number that determines the potential space for our thoughts. The space occupied by our thinking is (at the very least) the space of the possible combinations in which each synapse is active or not. And this number is two multiplied by itself not sixty-four times, as in the fable of the wise Persian, but hundreds of thousands of billions of times.

The resulting number is stratospheric; to write it down you would need thousands of billions of digits, 'so many, their number, more than the doubling on the squares'. Not even the most far-reaching cosmology deals with numbers this large. This number quantifies the immensity of thinkable space, of which we have yet to explore more than an infinitesimal corner. This is the boundless space opened up by combinations, by the art of combination, the *Ars magna*, the great art of Ramon Llull.

Are We Free?

Try lifting your index finger, freely deciding whether to raise the right one or the left. A second before you make your decision, nothing and no one can predict which finger you will raise, right? So would you think that your freedom had been diminished if you discovered that the outcome of your decision could in fact be predicted *before* you made your choice?

The study of nature teaches us that, at our scale, nothing happens without a cause, and that in general the world is deterministic, which is to say the future is determined by the state of preceding things. Therefore, a sufficiently precise observation of the state of the world could make it possible to predict even the outcomes of what we regard as free decisions. How can we reconcile this determinism with our experience of freedom of choice? The conflict between the necessity that we observe in nature and our feeling that we have freedom is the basic problem of free will.

The solution to that problem, which I think is the right one, is presented in one of the most beautiful pages of philosophy ever written: the Second Proposition, and its related *scoglio*, or commentary, in the Third Part of Baruch Spinoza's *Ethics*.

According to Spinoza, mind and body are not different entities. They are two ways of describing and conceiving the same entity, and each one is guided by necessity. What do we mean when we say that we make our own 'free' decisions? Spinoza's answer is simple and striking: it means that the outcome of the decision is determined by the complex inner working of our body/mind, and we ignore the complex causes that have led us to make the choice because we are not aware of the complexity of this inner working. 'Free will' is the name that we give to those actions of ours the causes of which depend of course on what happens in us, but which we are unaware of. Spinoza observes that the complexity of our body (today we would say of our brain) is very great. If we could know in sufficient detail how it functioned, we would see that, before a 'free' decision, there was already in place an unfolding chain of physical events that could have only one outcome.

Today, 350 years later, recent experiments in the field of neuroscience have led to an unexpected confirmation of Spinoza's idea, opening up a dense and fascinating dialogue between philosophers and neuroscientists.

An experiment in this area was conducted by John-Dylan Haynes at the Bernstein Centre for Computational Neuroscience in Berlin, the results of which have been published in the journal *Nature Neuroscience*. Haynes used functional magnetic resonance imaging, which is to say a scanner that 'photographs' the electrical activity of the brain, in order to observe the cerebral activity of different individuals in the process of making decisions. The subjects of the experiment had to freely decide to push a button on the left, or one on the right. The surprising result is that observation of cerebral activity *preceding* the moment of decision makes it possible to predict in advance the decision that will be taken. And this prediction can be made as early as several seconds before the actual decision! In other words, while you decide 'freely' whether to raise your left or your right finger, the decision is already predetermined, unknown to you, and at least several seconds before you think you are making it, by the bio-chemistry in your brain. What happens is precisely what Spinoza indicated: the sensation of making a conscious deci-sion appears to be nothing more than a psychological effect, subsequent to the biochemical events which have determined its result. Patrick Haggard, a neuroscientist at University Col-lege London, puts it like this in the latest issue of *Nature*: 'We think that we are choosing, but in reality we choose nothing.' I would not put it this way. I would say, rather, that what we call 'freedom to choose' is precisely the complex calculation that takes place in our brain. The outcome of a decision depends upon what is in our brain, that is, upon us.

The problems opened up by these experiments have to do not just with neuroscience, but with philosophy and ethics as well. For much contemporary philosophy, the problem of free will is no longer posed in terms of a Cartesian duality between body and mind, according to which the mind acts on the physical reality of the body by means of a special gland in the brain. Many philosophers today have little diffi-culty in accepting Spinoza's thesis on free will. But if free will is in this sense illusory, where does that leave individual responsibility? If someone commits a crime without having freely chosen to do so, should we therefore refrain from punishing the crime?

The answer, it seems to me, is obviously not. Using prisons and fines remains an effective practice for society, to defend itself from the actions of the individual in question, and to deter crimes by others, even (in fact, even more so!) in a deterministic world.

The important point, I believe, is the fact that the notion of free will remains useful precisely because we *don't* know the microscopic complexity that causes our behaviour. Our behaviour is in fact unpredictable, due to the complexity, as well as to the chaotic and even quantum aspects, of our bio-chemical make-up. The notion of free will, even if it is an approximate notion based on ignorance of potential causes, remains consequently the most efficacious when thinking about ourselves, just as Spinoza suggested.

But can any of us really accept, even when confronted with the evidence of a machine that foresees in advance which finger we will decide to raise, that our precious free will, if taken literally, is in the end just a sort of illusion? Or are we too attached to our pride as decision-makers, to the rhetoric of the freedom of the spirit, to ever accept such an idea? Spinoza himself, in the *Ethics*, suggests an answer:

> I hesitate to believe that men can be induced to reflect on all of this with equanimity, so firmly persuaded are they that it is only at the bidding of their minds that their body moves or stays still . . .

A Stupefying Story

This summer I found myself presenting my latest book at a dinner attended by the great and good of a section of London's culture industry: editors of the cultural supplements of newspapers, directors of museums, publishers, and so on. I was asked when it was that my curiosity was sparked to study the things that I write about in the book in question, *The Order of Time*. I hesitated briefly before deciding to tell the truth. I talked about the experiences I had with LSD when I was a teenager. I thought that I was probably taking a risk, since the subject is still somewhat taboo. The reaction I got was unexpected. One after another during the course of the dinner, people came up to me, smiling, happy to tell me about . . . the psychedelic journeys that *they* had gone on, forty years ago.

I was reminded of this when reading Agnese Codignola's fine book: *LSD: From Albert Hofmann to Steve Jobs, from Timothy Leary to Robin Carhart-Harris, the Story of a Stupefying Substance.* The title includes a pun: '*Stupefying*' means 'amazing', but also 'psychedelic'. The four names in the title sum up the recent history of psychedelic substances. Albert Hofmann was the Swiss chemist who first synthesized LSD and experienced its effects. In 2007 a global consulting firm placed him first on a list of 'living geniuses', together with the inventor of the internet. Timothy Leary is the Harvard psychologist who promoted the use of psychedelics to 'expand one's mind', becoming in the process one of the most noted and controversial figures of the counterculture of the sixties. Steve Jobs, the founder of Apple, currently the richest company on the planet, embodies the strong influence that psychedelics had on the world of high technology in Silicon Valley. Jobs has said that 'taking LSD was a profound experience, one of the most important things in my life', a phrase that I could echo. Robin Carhart-Harris is a young English neuroscientist who works at Imperial College and who recently obtained important results on the effects of psychedelic drugs upon the brain – results with implications for their potential therapeutic use – causing something of a stir and a renewal of scientific interest in this strange substance.

These four figures represent different phases in the saga of the particular class of drugs to which LSD belongs: psychedelics, or hallucinogens, on account of the intense and lively visual hallucinations that they produce. To the same class of drugs belong mescaline, psilocybin and similar substances present in mushrooms, cacti and other plants used in the religious rituals of various traditional cultures.

Agnese Codignola's book traces the arc of the impact of LSD on our culture. In the fifties the drug was promoted by celebrities. Cary Grant's enthusiasm for it remains notorious, the mainstream and clean-cut actor having claimed that it cured his depression. In Italy the drug was publicized passionately by no less than the American ambassador to the country, Clare Boothe Luce. When my parents, worried about their teenage son the 'drug addict', sent me to a psychiatrist in Verona to see if the drug had 'made me mad', the first thing that the good doctor said to me was that he also 'had tried LSD'. These were the years in which the Czech psychiatrist Stanislav Grof wrote that psychedelic substances 'could be for psychiatry what the microscope was for biology and the telescope for astronomy'. Then, at the beginning of the seventies, the use of psychedelics spread through youth culture, provoking alarmed and extreme reactions. Psychedelic substances were soon everywhere made illegal. Timothy Leary was given a thirty-year prison sentence. Scientific research into these substances was completely blocked, everywhere in the world.

Today, fifty years later, we are beginning to talk about the subject again, and the scientific world has realized that the suppression of such drugs has been excessive. If they are used with caution, in an appropriate environment, they have no known negative effects, and they are not addictive.

Caution is necessary because the psychedelic experience is extremely intense, and the negative effects that have occurred have been linked to their use in ill-adapted environments by people with existing mental problems. A vast inquiry conducted in the United States as part of the National Survey on Drug Use and Health has shown no increase in psychiatric problems among those who have used psychedelic drugs, when compared to the mental health of non-users.

Unfortunately, there are still alarmist anecdotes circulating on the supposed danger of these substances. It is common to hear, for instance, of suicides following the ingestion of LSD. But serious research has shown that the percentage of suicides does not in fact grow with the use of these substances. To make deductions from single episodes of suicide, as is still commonly done, regrettably, even by people in positions of responsibility, is like saying that *The Times* is dangerous because it has been found in the coat pocket of a suicide.

It is estimated that some 23 million Americans have used psilocybin, and no one has found any correlation between this use and addiction or toxicity. Recent research published in the *Lancet* on the dangers of taking chemical substances, for the individual and society, placed LSD in the very lowest rank, significantly below alcohol, tobacco, cannabis and many other substances used widely, including regularly in medicines. No rigid prohibitions exist for use of these higher-ranked substances, let alone for their use in scientific research. Given the indications of the potential of psychedelic drugs for therapeutic use, for problems ranging from depression to addiction to really dangerous drugs such as heroin, a number of different voices have been raised to say that the time has come to lift the taboo, and to grant at least that freer scientific research should be allowed.

Robin Carhart-Harris is one of the few who has managed to gain permission to study these forbidden drugs in recent years. In 2016 he caused a stir by publishing observations on the brains of subjects under the influence of LSD, obtained using techniques for recording images of cerebral activity. What can be seen is an explosion of activity. Substances such as LSD apparently act on the chemical connections between the neurons of the brain, allowing the awakening of new links. Many of these, though not all, are temporary. One hypothesis put forward is that residual effects that allow the reorganization of the cerebral structure provide the ground for therapeutic effects. These effects seem to be achievable with low amounts of the drug, or even with single doses.

Among the few serious hypotheses actually considered to explain something about consciousness is the theory of so-called 'integrated information', of which the neuroscientist Giulio Tononi is one of the principal originators. The theory suggests a correlation between the quantity of consciousness and the amount of integration of a structure elaborating information. The data gathered by Robin Carhart-Harris would seem to indicate, from this point of view, an effective state of 'augmented consciousness' which echoes curiously with the psychedelic ideology of the seventies.

A compelling description of a 'trip' – as the mental journey induced by such drugs came to be called – can be found in Nigel Lasmoir-Gordon's *Life is Just . . . Cambridge 1962*, a book that conveys the atmosphere of the dawn of a cultural revolution. But the most memorable literary description of the psychedelic experience is the classic one by Aldous Huxley in his novel *The Island*, in which a psychedelic drug called moksha is at the centre of a gentle, peaceable and wise culture on the island where he situates his Utopia.

A phrase frequently used to describe what happens during the intense and rich psychedelic experience is 'dissolving of the ego', or 'loss of a sense of self'. A psychedelic 'trip' can last for eight to ten hours, and many people, such as Steve Jobs, recall this experience as life-altering. In traditional cultures that use psychedelic substances for religious rituals, and for many young people in the sixties and seventies, the experience has taken on a mystical and religious connotation. The reason why psychedelic substances are frightening, according to Carhart-Harris in an interview in the *Independent*, 'is that they reveal aspects of the mind, and people are frightened of their own minds; they are frightened of the human condition'. I think the fear is due more to ignorance and prejudice.

If I was asked to try to sum up in a phrase what I think has stayed with me from those magical nights so many years ago, I would perhaps say that it was this: that the experience, for a number of hours, of a reality profoundly altered from our habitual perception of it, left me with a calm awareness of the prejudices of our rigid mental categories, and of the flexibility and potential depth of the inner world that our brain is capable of experiencing.

At that dinner in London, I became aware with increasing astonishment that my memory of such an experience was more widely shared than I had ever imagined. I don't know whether the majority of British people of my generation had 'acid trips', as they were called. Or whether those who did ended up directing museums and editing the cultural pages of newspapers . . . What seems certain is that LSD has had a lasting influence on a part of my generation. Perhaps now, after more than forty years of silence, we can begin to talk about it.

Why I am an Atheist

Various people have asked me why I don't believe in God. This is my answer.

Personally, I don't like people who behave well because they fear that otherwise they might end up in hell. I prefer those who behave well because they value good behaviour. I don't trust those who are good for the sake of pleasing God. I prefer those who are good because they actually are good. I don't like having to respect my fellow men and women because they are children of God. I prefer to respect people because they are beings who feel and suffer. I don't like those who devote themselves to others, and to justice, thinking that in this way they will please God. I like those who instead devote themselves to others because they feel love and compassion.

I've never liked feeling in communion with a group of people standing in silence inside a church listening to a service. I like to feel in communion with a group of friends; talking with them, looking into each other's eyes and at each other's smiles. I don't like emotion produced by nature because God has made it so beautiful. I like to feel moved by it because it *is* beautiful.

I don't like to feel consolation in the idea that I will be welcomed by God after my death. I like to look directly at the limited length of our lives, to learn to look at our sister, death, with serenity. I don't like shutting myself away in silence to pray to God. I like silence in order to listen to the infinite profundity of silence. I don't like to thank God: I like to wake in the morning, look at the sea and thank the wind, the waves, the sky, the fragrance of plants, the life that allows me to exist, the sun that rises.

I don't like those who explain to me that the world was created by God, because I believe that no one knows where the world came from; I think those who claim to know are deluding themselves. I much prefer looking the mystery in the face, feeling how tremendously moving it is, rather than attempting to explain it with fables. I don't like those whose belief in God gives them access to the Truth, because I believe that in reality they are as ignorant as I am. I think that the world is still a boundless mystery to us; I dislike those who have all the answers. I prefer those who are asking questions, and whose answer is: 'I don't really know.'

I don't like those who say they know what is good and what is evil because they belong to a Church that has monopolized God, blind to the many other different churches that exist in the world. There are so many different systems of morality in our world, all of them sincere. I dislike those who tell others what they should be doing because they feel that they have God on their side. I prefer those who make humble suggestions, who live in impressive ways that I can admire; who make choices that move me and make me think.

I like talking with friends, trying to help them when they are in trouble. I like talking to plants, and giving them water to drink when they need it. I like being in love. I like to gaze in silence at the sky. I like stars. I have an infinite liking for stars. I do not like those who seek refuge in the arms of religion when they feel lost, when they are suffering; I prefer those who accept that the wind blows, and that whereas the birds have their nests, the son of man has nowhere to lay his head.

And since I want to be like those who I like and admire, and not like those who I don't like and am not impressed by, I do not believe in God.

Hadza

It is still dark when we leave the lodge. I manoeuvre the Land Rover with excitement, following the half-spoken instructions of Hassan, our guide, who is still half asleep. The road is sketchy, and fording a stream causes me some problems. We leave the vehicle under a baobab and begin walking through the savannah. There are three of us: my partner, Hassan and myself.

After a long drive we catch sight of them. Five or six men sitting around a small fire. Hanging in a tree nearby, there are baboon skins, bows, a small musical instrument made of wood, an enormous python skin. A short distance away there is the circle of women, and less than a dozen very small huts. I crouch next to the fire, joining the circle of men. Most unusually for Africa, there are no greetings, but one of them offers me a piece of tree trunk to sit on and I understand I am welcome. He's a boy, with the darkest skin, an elongated skull, wide gentle eyes, a proud look, a baboon fur on his shoulders. Next to me a man is whittling an arrow with a blunt knife. I take out my Opinel, an outdoors sharp French penknife, and offer it to him. He tests the edge of the blade with his finger and laughs, then jokingly makes as if to cut a lock of hair from the head of the person next to him. Everyone laughs. He hands me back the knife, but I indicate with a look that he can keep it. Later on, I find out that his name is Sha-Kua. I stare into the fire along with the others and begin to feel a strange intoxication, a wild joy, an obscure feeling of having joined something primordial, a childhood game, something we have done as a species for hundreds of thousands of years, something for which we have evolved. In the company of these African men whose language I don't even speak, and who know so little about the world that I am from, I feel strangely at home.

They are Hadza, a population of hunter-gatherers. They live in a region of northern Tanzania. There are not many of them left. The migrations of Masai cattle-herders, and then the encroachment of the modern world, has significantly diminished their territory. In the seventies, the socialist government of Tanzania attempted to improve their standard of living by providing them with housing. The Hadza tried living in it for a while, but only before returning to the nomadic way of life that they preferred. I have heard about young Hadza who have gone to school, got good jobs – something precious in Africa, where hunger and poverty are common – but have decided to give it all up in order to return to a life of hunting. Sitting with these men around the fire, it is a little easier for me to understand why.

A little easier still when, shortly afterwards, we leave for the hunt. We walk silently and warily in the savannah, hands tensed on bows. The men spread out, remaining in contact through a series of quiet whistles that I can hardly distinguish from the birdsong. Sha-Kua hits a dik-dik, a small antelope, with one of his arrows. We follow the trail of its blood and find the poor animal, pierced through by the arrow, in a bush where it has hidden itself to die. The men light a fire by rubbing wood together, as easily as I would with a box of matches. I have a go at it myself, without success; the boy laughs and begins to teach me. The little antelope is put on the fire, and we all eat together. Sha-Kua has removed one of its horns as a gift for me.

Have I succumbed to a kind of naïve romanticism? Or to that infinite capacity of ours to project our own thoughts and fantasies on to others? I don't know, but I do know that my heart was still racing even as we returned in single file towards the village. One of the men carries on his shoulders the remaining half of the antelope, for the women who in the meantime have been searching for fruits, berries and roots. I feel like a child invited to join in playing the most wonderful game. A part of me would like to stay with these men who laugh, joke, teach me things, walk with bare feet in the savannah, calmly and proudly, with bows in their hands. Isn't this what we were born to do? Isn't this what we have always done, for countless millennia? A few friends around a fire, a hunting expedition, the return home to the women. Back in camp, next to the fire again, a pipe is passed round, and this time I decide to take its acrid smoke into my lungs. It is a kind of mild marijuana that grows naturally in the area.

Daudi Peterson, an American anthropologist who grew up in Tanzania and has lived extensively with these people, has collected in a beautiful book with the wonderful title *Hadzabe: By the Light of a Million Fires*, stories and images by Hadza speaking in the first person about their lives and their view of the world. They live in small independent groups in which decisions are taken in common, with the women having as much say as the men. When two young people fall in love, the man goes hunting for a baboon and gives it to the father of the woman as a sign of gratitude, and the two lovers start to live together. The entire group takes care of the children. Elders are respected, their stories are listened to around the fire, but they have no more influence in decision-making than anyone else. There are no social classes and there is no hierarchy. There are no leaders. Anyone who thinks themselves superior is mocked. Anyone in disagreement with the group, or unhappy in a given situation, can go their own way. There is no property; food is immediately divided up and distributed, since the meat and other components of their diet cannot be kept. Today anthropology teaches us that this is the way our species has lived for hundreds of thousands of years, an immensely long period of time. Cultivating fields, keeping cattle, building cities, reading books, erecting temples and cathedrals, surfing the internet, are all incredibly recent innovations by comparison. Perhaps we are not really used to the novelty of these things yet, to the discontents of civilization?

And the Hadza? They are convinced, like the rest of humanity – whether Chinese, British or Veronese – that their own way of life is the only reasonable one, and that all the others are strange. They observe that the tribes in the area who live from cattle-herding or agriculture suffer from hunger and even famine (a few years ago a drought decimated the cattle of the Masai, reducing the population to utter poverty). The Hadza do not know famine: there is never any lack of animals to hunt or fruits to gather in the savannah. I tried to get Hassan, our guide, to ask Sha-Kua what he thought about us. Hassan was born in a village near to Hadza territory, and has known them since he was a boy. He tells me how he used to find animals struck by Hadza arrows, and used to seek them out to return them. He is on friendly terms with the Hadza. But his answer to my question sounds nonsensical to me: 'He thinks that you are interested in them because they are such good hunters and you want to learn from them.' I wonder if he is pulling my leg. But then I ask myself whether, as boys in Verona in the seventies, we had the slightest interest in the American tourists we would see passing by. Perhaps Sha-Kua had a similar lack of interest in how 'the others' lived. Like the majority of men, he cared about his friends, his hunting, his woman. And perhaps it was this that had allowed his people to remain so uninfluenced and unchanged for centuries, and to keep living as the fathers of their fathers had lived. Perhaps Sha-Kua and his people are less gnawed by curiosity, by the desire to know more. Perhaps it was this that pushed some of us out of Africa, spread us across the planet, made us domesticate animals and plants, investigate the stars, ask a thousand questions, build villages, cities, metropolises and megalopolises. Perhaps those who let the Neolithic revolution pass them by – along with all the other, lesser ones – simply had less curiosity, less desire to look beyond the next hill. Or perhaps they were far-sighted enough to see the risks of imbalance that such changes brought with them. I don't know.

So here we are, you and I, Sha-Kua, looking at each other across the millennia since our ancestors took such different paths, and it seems to me that I can see in your eyes both the value and the cost of the paths taken. I go around the world, read books, have health care if I fall ill; you don't. I have an irremediable restlessness and can't stay still. As for you, I don't know. But the things that matter to us have remained basically the same, and in my naïve imagination you possess them all, because our biology has evolved to do what you do, and not what I do. I certainly wouldn't know how to live like you any longer. And it makes no sense to wonder, however irresistibly, if this other path was worth taking after all. We are what we are.

I'm leaving you my French knife. It's of much more use to you than to me. You're giving me this small antelope horn. For me it is a reminder of a life that's been lived for one hundred thousand years. A lost life that you, Sha-Kua, are one of the very last to suffer, or to savour.

A Day in Africa

Today I have decided to leave the comfort of the Institute of Mathematics in Mbour, where I have been spending a few weeks, in order to venture out to see something of the 'real' Africa. I hail a collective taxi in the street, squeeze my diminutive self between a couple of well-built ladies in tight-fitting colourful clothing, and reach the centre of Mbour with a ride costing a hundred African francs – the equivalent of less than fifteen European cents. Before leaving the coast, I take the opportunity to have a look at the market. It is much bigger than I expected: the place is swarming, pungent, colourful and grimy, covering a seemingly endless quarter and becoming increasingly dense the closer it gets to the beach, where dozens of small fishing boats unload hundredweights of fish that are carried off in every direction. With some effort I manage to extricate myself from this unsmiling, toiling sea of humanity and let another taxi take me to the only tarmac'd crossroads in Mbour: the one where Route Nationale 1 branches off from the coastal road and heads towards Mali. I'm aiming for the village of Sandiara, twenty kilometres inland.

After a few negotiations I find a car willing to take me there for one thousand francs; less than two euros. The landscape turns out to be dreary savannah, dotted with baobab trees, and on arrival I find that Sandiara is more like a small town. A large group of people is clustering around something. I approach discreetly and manage to catch a glimpse of what is at the centre of the crowd. It's a man sitting on the ground, covered from head to foot in mud and dust. He seems distraught, desperate even. His hands are tied behind his back, and his feet are bound. His eyes are fixed on the ground. The crowd surrounding him is noisy, commenting loudly as it watches him. A young man informs me that he is a madman. Then quickly corrects himself: he is a 'killer'. A few particulars emerge: he has knifed someone. I ask what will happen now. Now he will be taken to the next village. 'Now' in Africa, it seems to me, is an imprecise term that can be more accurately understood to mean 'sooner or later'. There is no one here in uniform, just the small crowd watching and commenting. Nothing happens. I feel pity for the man. He seems more than just desperate: annihilated, rather. As if he had yielded completely to this crowd and the way they were looking at him. I find myself thinking that I am the only white man for twenty kilometres and that, as a stranger here, there is not much that I can do. I wander for a while in the sandy streets of the village, observe the children playing, the blacksmiths, the small mosque, the dirt that covers everything – then return to the road and find a bus that takes me to the next village, Tiadiay. I buy some bread from one of the countless women vendors who teem in every street, and head in the direction that I've been told will lead to Sao.

I chose Sao for its name, because I liked the sound of it. I saw it on the map. It was away from the major roads, but not too far away. As I walk towards the end of the village, a sweaty-faced man in a yellow robe asks me where I am going, and I tell him that I'm going to Sao. Generally speaking, I'm wary of anyone who approaches me, and especially of those with sweaty faces, but it simply won't do here to be too stand-offish. '*Sao?*' 'Yes, Sao.' He offers to take me there for three thousand francs. I suggest two thousand, and he gestures for me to follow him to his car. It is an incredibly old yellowish Peugeot, even more dilapidated than the wrecked bangers of Mbour. One of its doors doesn't close, and for half the journey Barri (I have discovered that he is called Barri) holds it shut with his arm. The other half he spends attempting to close it by opening and slamming it shut again. In vain. After covering a few kilometres, he slows down, pulls over and tells me that we have to take a barely visible track in the sand, to our left. I don't say anything, even if I have a sudden moment of apprehension. Barri hardly says anything either, and I do not like this. He answers only with monosyllabic non sequiturs. In an attempt to make conversation, I had pointed to the clouds and asked if it was unusual to see such a sky in Senegal in January. He replied: 'the sky'. He does not seem very sharp, and I find this reassuring.

And then we arrive in Sao. A Sao that turns out to be completely unlike the one I was expecting. I'd expected another teeming village black with dirt. Instead it is a half-empty, scattered affair, consisting for the most part of huts studding the savannah between the baobabs, suffused by the golden hue of the sand and the straw. As soon as I get out of the car a band of wide-eyed children rushes up, as if a flying saucer had just landed. An old man appears, a few women. They can't understand what I'm looking for. I try to explain that I'm curious, that I would just like to look around the village, if they don't mind. This strikes them as pretty strange. They offer to accompany me, to guide me. The old man beckons an extremely beautiful young woman and tells me that she will be my guide. If it wasn't for the puritanism of Islam, perhaps this would have seemed to me like some kind of ambiguous offer. As it turns out, what I needed more than a guide was someone to keep away the crowd that had formed. A small man in festive mode pops up with a drum which he beats like a madman, and everyone breaks into laughter and claps their hands. A young woman begins to dance.

They tell me they are pounding the millet, taking for granted my knowledge of the fact that the village exists thanks to millet, and that I know everything about how it is cultivated and prepared. They take me to see the women who pound it with enormous wooden pestles in equally out-size wooden vessels. They are the same kind of pestles you find throughout Africa, but every time I have seen one being used it has been working to grind a different substance. I ask how many people live in the village and am told that I should enquire at the school. Great, a school! I ask to be taken there, and Barri, together with a muscular, kindly young man who had been following us, guides us across the sand, past the goats and the baobabs, towards the school. It isn't far to go. It consists of a few huts and sand-coloured walls. We go to see the head, who immediately busies himself dusting a chair for me to sit on. He is an intelligent man, passionate, devoted to his school, lively and engaging. He tells me about the educational programmes that fall here from above – the latest from Canada, no less – about the teaching of Arabic and of religion, of the many challenges but also of the desire to study that all of the boys have, and all of the girls too, he is keen to emphasize. The environment here is good, and Africa is like this, he says with a smile: always disastrous, but always elated. We give only a passing glance at the children 'who sometimes don't pay attention because they don't get enough to eat at home'. He speaks with humility, but with awareness too of the crucial importance of what he and the other four teachers are trying to do for these children. I would like to ask him more about the teaching of Islam in primary schools, but fear that this would be a sensitive subject. He shows me on the timetable the hours given to the teaching of Arabic and religion: one hour per week, more or less. 'Are there any Christian children?' 'Yes, a few,' he tells me: during the hours devoted to the teaching of Islam, they leave the room. Just like in Italy, only the other way round: in Italy the Muslim children leave the room when the subject is Christianity. My heart sinks at the thought of the stupidity of humankind, but I prefer to avoid the subject. I begin to take my leave, and thank him sincerely; he is visibly pleased by our meeting. Before leaving I mention that I would like to contribute something, to help with buying materials for the school – exercise books, pens, and so on – and ask if I can leave it in euros. I produce a worthwhile sum; he promptly calls one of his assistants so that he can be witnessed receiving it. We part so cordially, almost emotionally, it seems to me. Even if I'm not sure exactly why.

Barri, with more forethought than I'd had, has not gone anywhere. It's hard to see how I would have left otherwise, given that the only other form of transport I'd seen in this village half lost in the savannah was an ancient-looking donkey. I suggest that he should drive me north, up to Route Nationale 2, the road that goes towards Mauritania. From there I should be able to return to Mbour with public transport. We haggle for a while before agreeing on a reasonable fare. We head off with Barri holding the door shut with his hand. The route is long, on a dusty and sun-baked dirt track. The car seems to be made of only encrusted sand, rust and vestiges of antique-looking plastic. And yet despite this, between arid expanses and isolated, desolate villages, it continues to move.

There are no other cars. I watch the country sweeping past, through the wide-open window from which the glass has long since gone. It occurs to me that the majority of human-kind lives more or less like these men and women, like these dust-covered children – and not at all as I do. We are the exceptions, sequestered and well defended in our gardens of wealth and hygiene.

A few hours later we arrive at Khombole, and I see again the dirt that accumulates in roadside villages along so many of the main roads in African countries. Here it is particularly bad. I wonder if there is any relation to the fact that France is the country that, as they frequently say here, 'colonized us'. I lack the courage to eat any elaborate food, and make do instead with oranges, bananas and bread. I look for somewhere to eat in the shade and alone, but my solitude is short-lived: in no time at all I am surrounded by a band of children. I play with them, taking photographs and displaying them on the camera screen. The young girls smile coquettishly; the boys laugh loudly and show off. I make the mistake of giving them some biscuits, and am forced to retreat when they fall over each other to grab some more . . .

I catch sight of a broken-down overcrowded bus that is heading in the right direction and board it. I arrive at Thies quite late in the day and realize that that I need to hurry if I am to avoid getting back in the middle of the night. A kindly old man wearing a long white tunic accompanies me to the *gare routière*, where I ask if there is a bus to Mbour. There is one. All I have to do is sit and wait to see if anyone else is going to turn up wanting to go to Mbour. This is how so much of public transport works in Africa. You wait. Perhaps for hours. Sitting on the bus, or a stone, in the midst of unspeakable rubbish and flies, in the *gare routière*. Half a continent spends an inordinate number of hours just waiting. I take advantage of the time to read. I've brought with me a short book I found in the only shop in the region where there was food that did not seem coated in dirt. It tells the story of a young Senegalese educated in a Koranic school before the arrival of European teaching, who is subsequently sent to a French school and eventually ends up in Paris, studying philosophy at the Sorbonne. It is a sad story, of the hesitation between different worlds, of the estrangement of being African in a Western global culture – or perhaps, ultimately, of the estrangement of being human. When the bus finally starts, after hours of waiting, I am far into the book and am perceiving the country around me from its disquieting point of view. I look at the savannah rushing past the open window. Nearby, huts; in the distance, in the smoke, the outline of a factory.

It is dark when we arrive in Mbour. Mbour is the metropolis; after a day spent in the vastness of the interior, it seems almost Dante-esque. There is manic traffic filling the only surfaced road. Clouds of dust are illuminated by car headlights. It is thronging with noises, darkness and light, confusion, the haunted eyes of passers-by; it is like the antechamber of an inferno. The bus arrives in the *gare routière*. I get down, buy some oranges and realize that the price has doubled due to the colour of my skin, but am not particularly bothered. Then I become aware that the *gare routière* is just behind the big candy-pink mosque that I had glimpsed a couple of times in passing. It had seemed to have the air of being self-enclosed, unreachable. When I had mentioned the mosque to the owner of the restaurant I sometimes went to, the only white man I had met in the area, and asked him if it was possible to visit it, he mumbled a half-hearted 'no'. But now there were people coming out after evening prayer. I decided to take a chance and try to go in. The worst that could happen is that I would be refused admission.

There is a thin chain delimiting the area occupied by the mosque, and on the other side of this chain there is something resembling calm. I reach the railing. The people coming out are putting on their shoes. I take off my grimy sandals and carry them with me across the area. I feel the soft rug of fake grass beneath my feet. The faithful are coming out in groups, as happens in European churches. Except that here they are all men – and nearly all of them of a certain age, or very old. I'm surprised that they have such an immaculate, dignified, serene air about them. When they pass, they greet me. Many of them smile. In this country people don't smile much, but it seems that here they do. I wonder what I must look like to them. I'm quite obviously dirty after a day spent travelling, and my arms are bare, whereas everyone here has long sleeves. I'm carrying a rucksack and wearing a basic straw hat, conspicuously not dressed to enter such a place as this. And I am white-skinned: so white, in contrast to everyone else, as to actually shed light. But they are smiling at me, nodding in a kindly way. Apparently, they are pleased to see me going to the mosque. I had feared that I would be unwelcome, or regarded with hostility . . . I arrive at the door. Cautiously, bare-footed, I step inside and look around. A young man is hurrying towards me with a worried expression on his face. He says something that I can't understand. It is clear that I have overstepped the mark in some way. He points to the sandals that I'm carrying and I realize what it is: the rule is not that you shouldn't wear shoes in the mosque, it's that shoes shouldn't enter it. I quickly step back through the door and leave the shoes with the others. I'm about to go back in when an old man approaches, gives me an encouraging smile and says something to the young one who had stopped me. He takes my shoes, places them in a black plastic bag and carries them into the mosque himself, then hands them back to me. With embarrassment I try to explain that I'm not worried that they'll be stolen, that I'm quite happy to leave them outside . . . but he smiles, and the young man smiles as well. In response I pick up my shoes, thank the two men with my eyes and carry on into the interior of the mosque. I am speechless: there are places in the world where rules are less important than kindness.

Nearly everyone has left by now. There are a few still lingering, but the space is so vast that it gives the impression of a great void. Of a great calmness. Of a great silence. I sit on the floor, on rugs, and lean back against a wall. The contrast with what's outside could hardly be greater. Outside it is infernal; inside we are in paradise. Everything is spotless, impeccably clean. The walls and columns are painted a shining, clear white. The long, simple, elegant, inviting carpets are patterned with a dignified green and black arabesque. They are arranged in parallel, regular rows. The light is diffuse but clear. The arches and columns lift one's gaze and heart upwards. The few people still inside are not speaking quietly, as they do in churches, but are talking to each other normally – but their tone of voice is calm, one might even say noble. There are no furnishings, glitz, ostentation of wealth, images of agony on the cross, candles, obscurity, old paintings of ecstatic faces, gold leaf. There is just a large, serene space. A welcoming space. Something human, terribly human, where the core of being human seems to be to gravitate towards the essential, the absolute.

And suddenly it seems as if I have glimpsed for a moment

something of the heart of this place. This labouring, impoverished, dusty, chaotic Africa conceals within itself, in a place that seems to me the most inaccessible, the calm dignity of these men, the wonder of this perfect space offered to man so that he can be fully himself and at peace with himself, as I have never found anywhere else. Profoundly at peace. And for a moment, despite being a fully-fledged atheist with no hesitation, I feel that I understand what it means for so many people to abandon themselves to the total omnipotence of a God who is not a father, but is the true and complete Absolute.

I leave with my own sense of serenity. Perhaps these are simply physical reactions to the day's heat, travelling, dehydration, encounters, stress and general fatigue. Or perhaps I have actually learned something, one small additional thing, about the complexity of being human.

The Festive Season is Over

The festive season is over. For another year, the wave of emotions has passed, of lunches, cakes, relatives, little trips, presents, and all those other things that go with Christmas. I'm always surprised at how everything is affected by this season. And how, even if you try to resist, you end up being pulled along by it.

You can't not go to see a dear relative. You can't turn up without at least a small present. You can't avoid dressing the table up a little; putting up a small tree, a basic nativity scene, some coloured lights or a candle. You have to mark this time of year in some way. Where does it come from, this immense push towards festivity that we all feel?

For Christians, obviously, Christmas is the celebration of the birth of the Saviour. The celebration of the arrival of Him who saved us. It is a moving commemoration: the arrival of the invisible into the world. The nativity scene re-creates this magical moment, bathed in the light of our emotions.

But festivities at the end of December are a great deal older and deeper than Christianity, which appropriated them, embedding its own mythology and theology there and displacing other traditions. Traces survive, transformed: in ancient Rome it was customary to light candles and to exchange presents towards the end of December, long before the birth of Christ. The pagan tribes of the north celebrated the winter solstice long before the Christian message reached them. The force that compels our behaviour towards the end of the year is more ancient than Christianity. What kind of force is it?

A great book, *Ritual and Religion in the Making of Humanity*, published a few years ago by one of the major anthropologists of the twentieth century, Roy Rappaport, is dedicated to the ancient origins of rites and customs. Rites, if you think about it, are something strange and difficult to decipher from the perspective of naïve modernity. A rite is a gesture, an action, a word that is repeated in the same way, more or less regularly, and which carries an intense emotional charge for whoever performs it, even when it doesn't seem to have any direct utility. Or at any rate not the kind of utility that would justify the extraordinary power invested in it.

Why, for millennia, have we exchanged gifts at the end of December? Empires have collapsed; entire populations have been slaughtered; we have changed religion several times; we have been rich and poor, dominated and domineering; we have believed in witches and have gone to the moon – and with absolute regularity, at the end of December, we have exchanged a small gift and lit a candle or a small light. Is this not remarkable?

According to Rappaport, the birth of rituals can be traced to the very formation of humanity itself, to the emergence of that articulate language that today is so markedly characteristic of our species. Rituals, according to Rappaport, have a key function in constructing us as human beings, particularly as social beings.

Ritual behaviour, which is to say elaborate, complex, repeated gestures without an obviously direct objective, are common in many species of animals, and often anchor the formation of durable ties – as in the intricate courtship rituals of many monogamous species. Among our own species, language has driven the construction of a multi-layered abstract world, where new entities come to life that did not exist before (laws, marriages, sentences, contracts, kingdoms, nations, property, rights . . .), and which have both a status in reality that determines our actions, and power as components of a shared system, adhesion to which follows deep-seated laws in our mental structure as primates that are shaped by ritual gestures and regularly reinforced by ritual gestures. Ritual, in short, is the foundation of the elaborate human social and spiritual reality within which a large part of our lives develops.

So the common life of two individuals who love each other is supported by the ritual of marriage; the professional life of a doctor is supported by the ritual of qualification to practice; prison time is supported by the ritual of court appearances and trial; the legitimacy of a parliament rests upon an electoral ritual; the validity of ownership of my home depends on a ritual involving solicitors; the inner life of a Catholic rests upon the weekly ritual of Mass, and the interior life of a Buddhist upon the ritual of meditation. The scientific life of my small research group in Marseille depends on our somewhat thin ritual of more or less regular meetings over a sandwich to talk about physics . . . And so on, endlessly.

We use the repetition of structured and regulated gestures to put order into the chaotic flow of reality, and to give us points of reference – points where we can anchor our perception of and our being in the world.

I'm not sure if Rappaport's reading of rituals is right when looked at closely, in all its details. I don't know either the extent to which it is supported by contemporary specialists in the field. But it teaches us something important and profound: as human beings we are multidimensional; we are made of various strata that we don't in general completely understand. What's more, if we aren't involved in actually studying them, we are governed by rules that we may not even be aware of. We give them names, and let ourselves be carried away by them and by life.

And every Christmas, whether we are fervent Catholics or outright atheists, we go home to see our ageing parents, and exchange presents with our friends. And so the year and the world turns in an orderly fashion: we are reassured that the ties of affection hold us together, we feel at home in the world. We are ready for life to begin again.

This Short Life Feels Beautiful to Us, Now More than Ever

The development of the epidemic in the UK is tracking what happened in Italy, a couple of weeks behind. Responses follow the same path: incredulity at first, fear, difficulty in accepting the reality, reaction times often far too slow.

It is difficult to accept the human fragility that this crisis exposes. We humans are not as powerful as we might have thought. In rich countries, we used to see the worst disasters happening elsewhere. In Italy, we watched China with a sense of security: it wouldn't get to us. When the virus took hold, we watched other countries, including the UK, making the same mistakes. Only a few weeks ago, I heard an American saying on television: 'We are the most powerful country in the world, the epidemic will do nothing to us.' He no longer says so. It is a humbling experience for everyone.

It is not really anyone's fault. It is not like war, triggered by human folly. There have been errors and negligence, of course, and we are certainly making more mistakes, but taking decisions in such unusual situations is difficult; we do what we can, groping in the dark. Next time we will be better prepared, react faster, listen better to the warnings from science. But it is a weak and poor reaction to search for somebody to blame – politicians who could have woken up to the crisis earlier; China, who could have sounded an earlier alarm; governments, who could have been better prepared.

The reality is that this disaster has no culprits. Simply, in spite of our hubris, we are still in the hands of nature. Sometimes she showers us with gifts; sometimes she mistreats us brutally, with sovereign indifference. Science and knowledge are the best tools we have. They allow us to avoid errors, such as those we made in the Middle Ages with processions to ward off the plague, which instead spread the infection. But never has it been so clear that science cannot solve all problems. Our splendid intelligence surrenders to a small virus which is little more than a speck of dust. Science is the best tool we have found, let's hold it dear; but we remain fragile when faced with a powerful and indifferent nature.

Our Western arrogance is being put to a hard test today as well. Italy, which prided itself on one of the best health systems on the planet, has received help from Cuba, China, Russia, even from Albania. Weren't these the countries we used to say were on the wrong path? The countries that have so far better defended themselves are Singapore, Hong Kong, Taiwan, Korea, China itself. Weren't we Westerners the best in the class? When this has passed, it will be time to review some of our presumptions.

It will pass. All the epidemics of the past have passed. Nobody is yet clear about the effect this will have on our lives, how disruptive it will be, how much it will cost each of us. Perhaps we will review some assumptions about the free market: even the most strenuous defenders of the total freedom of the market today cry out: 'The State should help us!' In times of difficulty, it becomes clear that collaborating is better than competing. My secret hope is that this will be our conclusion from the current crisis. Problems are best solved together. Humankind can survive only if we work together.

As the Western country unwillingly leading the way through

these difficult times, this, I think, is the humbling lesson Italy is learning. For now, we are struggling together to earn a little more life for our loved ones and ourselves. Because this is what we are doing: helping our doctors do what they are capable of doing – buying us days, years, more life. This is not a natural right of ours. It is a privilege that we have gained gradually, through collaboration and accumulated knowledge, thanks to civilization.

More than 22,000 Italians have died in the last two months; more will die, taken away by the epidemic. These are terrible numbers. But many more people die every day without the epidemic, any year, any week. The sorrow of losing a loved one is profound. But it has not been initiated by the epidemic: it has always been with us, and always will be. Twenty-two thousand deaths are a lot, but that number is far lower than the deaths each year from cancer. Or from heart disease. Or simply from old age. And it is immensely fewer than the number of deaths in the world from hunger or malnutrition. What this epidemic is really doing is putting in front of our eyes something that we usually prefer not to look at: the brevity and fragility of our life.

We are not the masters of the world, we are not immortal; we are, as we have always been, like leaves in the autumn wind. We are not waging a battle against death. That battle we must inevitably lose, as death prevails anyway. What we are doing is struggling, together, to buy one another more days on Earth. For this short life, despite everything, seems beautiful to us, now more than ever.

Image Credits

Seven Brief Lessons on Physics

'With the publication of this million-selling book, Rovelli took his place with Stephen Hawking and Richard Feynman as one of the great popularisers of modern theoretical physics' *Spectator*

These seven short lessons guide us, with simplicity and clarity, through the scientific revolution that shook physics in the twentieth century and still continues to shake us today. In his enlightening overview, Carlo Rovelli explains Einstein's theory of general relativity, quantum mechanics, black holes, the mysterious architecture of the universe, elementary particles, gravity, and the nature of the mind.

Reality Is Not What It Seems

'May genuinely alter how you see the world' *The Times*

In this mind-expanding book, Carlo Rovelli shows how our understanding of reality has changed throughout centuries, from Democritus to loop quantum gravity. Taking us on a wondrous journey, he invites us to imagine a whole new world where black holes are waiting to explode, spacetime is made up of grains, and infinity does not exist – a vast universe still largely undiscovered.

The Order of Time

'A captivating, fascinating and profoundly beautiful book' John Banville, *The Irish Times*

From Boltzmann to quantum theory, from Einstein to loop quantum gravity, our understanding of time has been undergoing radical transformations. Time flows at a different speed in different places, the past and the future differ far less than we might think, and the very notion of the present evaporates in the vast universe. With his extraordinary charm and sense of wonder, bringing together science, philosophy and art, Carlo Rovelli unravels time's mysteries...